Springer Texts in Statistics

Series Editors

G. Allen, Rice University, Department of Statistics, Houston, TX, USA

R. De Veaux, Department of Mathematics and Statistics, Williams College, Williamstown, MA, USA

R. Nugent, Department of Statistics, Carnegie Mellon University, Pittsburgh, PA, USA

Springer Texts in Statistics (STS) includes advanced textbooks from 3rd- to 4th-year undergraduate courses to 1st- to 2nd-year graduate courses. Exercise sets should be included. The series editors are currently Genevera I. Allen, Richard D. De Veaux, and Rebecca Nugent. Stephen Fienberg, George Casella, and Ingram Olkin were editors of the series for many years.

Silvia Bozza • Franco Taroni • Alex Biedermann

Bayes Factors for Forensic Decision Analyses with R

 Springer

Silvia Bozza
Department of Economics
Ca' Foscari University of Venice
Venice, Italy

Faculty of Law, Criminal Justice and Public
Administration, School of Criminal Justice
University of Lausanne
Lausanne-Dorigny, Switzerland

Franco Taroni
Faculty of Law, Criminal Justice and Public
Administration, School of Criminal Justice
University of Lausanne
Lausanne-Dorigny, Switzerland

Alex Biedermann
Faculty of Law, Criminal Justice and Public
Administration, School of Criminal Justice
University of Lausanne
Lausanne-Dorigny, Switzerland

Published with the support of the Swiss National Science Foundation (Grant no. 10BP12_208532/1)

ISSN 1431-875X   ISSN 2197-4136   (electronic)
Springer Texts in Statistics
ISBN 978-3-031-09838-3   ISBN 978-3-031-09839-0   (eBook)
https://doi.org/10.1007/978-3-031-09839-0

This Springer imprint is published by the registered company Springer Nature Switzerland AG
The registered company address is: Gewerbestrasse 11, 6330 Cham, Switzerland

To our families

Preface

The introduction of scientific evidence in legal proceedings raises a host of intricate questions and themes, ranging from the architecture of legal systems across contemporary jurisdictions and psychological aspects of judgment and decision-making, to principles and methods of logical reasoning and decision-making under uncertainty. Over decades of theoretical and practice-oriented research, scholars in fields such as law, statistics, history, philosophy of science, psychology, and forensic science have come to the understanding that the sound use of scientific findings in evidence and proof processes critically depends on the ability of forensic scientists to use formal methods of reasoning, so as to ensure a coherent approach to dealing with and communicating about uncertainty. The focal point of these developments is the recognition of probability as the reference method for measuring uncertainty.

It is thus hardly surprising that, in recent years, the intersection between law and forensic science has seen an increase in the number of reports, guidelines, and recommendations issued by eminent societies, review panels, and expert groups that insist on the importance of aligning the interpretation of scientific evidence by forensic scientists to a probabilistic measure of the value of evidence.[1] This measure is the likelihood ratio and has been widely described in peer-reviewed articles and textbooks.

What is less often recognized, however, is that the likelihood ratio is merely a particular instance of a more general concept, known as the *Bayes factor*. While the likelihood ratio is typically presented in the focused context of evidence-based discrimination between pairs of competing propositions, the Bayes factor is a method of choice for approaching a more comprehensive collection of problems commonly associated with the use of measurements and data in forensic science.

---

[1] Examples include documents issued by the Royal Statistical Society (Aitken et al., 2010), The Royal Society of Edinburgh (Nic Daéid et al., 2020), The UK Forensic Science Regulator (Tully, 2021), The European Network of Forensic Science Institutes (Willis et al., 2015), The Association of Forensic Science Providers (Association of Forensic Science Providers, 2009), and expert communities, in particular sub-fields of forensic science, such as forensic genetics (e.g. Gill et al., 2018) or forensic voice comparison (Drygajlo et al., 2015; Morrison et al., 2021).

Examples include the comparison of probabilistic models, model selection, and decision-making regarding competing theories and model parameters. We believe that by becoming acquainted with Bayes factors across a range of different applications, forensic scientists can strengthen the use of probabilistic methods in their respective disciplines. Forensic scientists should also gain an understanding of the role of Bayes factors in coherent decision-making under uncertainty. The core idea of this book on Bayes factors, the first on this theme in forensic science, is to address these questions.

Bayes Factors for Forensic Decision Analyses with R is a new Bayesian modeling book that provides a self-contained account of essential elements of computational Bayesian statistics using R, a leading programming language and a freely available software environment for statistical computing. This book features a well-rounded approach to three naturally interrelated topics. The first is probabilistic inference. As a core concept of Bayesian inferential statistics, Bayes factors are ideally suited to help forensic scientists think about the logical and balanced evaluation of the value of evidence. This is a necessary preliminary to coherent reporting on scientific evidence. Second, this book highlights the logical connection between probabilistic reasoning, using Bayes factors, and decision analysis under uncertainty. This perspective involves the decision-theoretic (re-)conceptualization of questions that, in classical statistics, are often framed as problems of hypothesis testing using a disparate set of concepts, such as p-values, that have a longstanding and well-documented history of misinterpretations by both scientists and recipients of expert information. Here, Bayes factors provide a sound and defensible alternative. The third theme that this book covers is operational relevance. Thus, throughout this book, all key concepts are systematically illustrated with hands-on examples and complete template code in R, including sensitivity analyses and explanations on how to interpret results in context. This usefully complements the theoretical and philosophical justifications for the coherent approach to inference and decision emphasized throughout this book.

Besides explaining the role of the Bayes factor as a guide to reasoning and as a preliminary to coherent decision analysis, the original contribution of this book is to work out the relevance of these topics with respect to two main forensic areas of application: investigation and evaluation. The first, investigation, refers to discriminating between general propositions of interest, i.e., when no named person (or object) is available for comparative examinations with a given trace, mark, or impression of unknown source. The second, evaluation, is concerned with assessing the meaning of evidence with respect to specific propositions of interest, e.g., whether given trace material, a mark, or an impression comes from a particular person (or object), rather than from an unknown person (or object). While investigation and evaluation pertain to distinct procedural phases with specific needs and constraints, they involve inferential and decisional tasks that have common conceptual underpinnings that can be formally captured, analyzed, and expressed in terms of Bayes factors, and embedded in a coherent framework for decision analysis.

This book does not contain recipes nor does it intend to prescribe what scientists should do. Instead, the aim of this book is to provide forensic scientists with a sound analytical framework for inference and decision analysis that allows them to critically rethink their current approaches drawn from more traditional courses in probability and statistics. As prerequisites, readers should have a minimal background in probability and statistics including, ideally, notions from Bayesian statistics. With its balanced presentation of theoretical and philosophical background, together with practical illustrations, this concise book seeks to make an original contribution to forensic science literature. It will be of equal interest to forensic practitioners and applied forensic statisticians, and can be used to support courses on Bayesian statistics for forensic scientists. Occasionally, we will refer to datasets and computational routines, available as online supplementary materials on the book's website at http://link.springer.com/.

This book presents materials developed through a longstanding collaboration between the authors. Their research was supported, at various instances, by the *Swiss National Science Foundation*, the *Foundation for the University of Lausanne* (Fondation pour l'Université de Lausanne), the *Vaud Academic Society* (Société Académique Vaudoise), the *Department of Economics of Ca' Foscari University of Venice*, and the *School of Criminal Justice of the University of Lausanne*. The authors are deeply indebted to Colin Aitken and Daniel Ramos for their valuable advice, to Lorenzo Gaborini for sharing routines developed in his Ph.D thesis, and to Luc Besson, Jacques Linden, Raymond Marquis, Valentin Scherz, and Matthieu Schmittbuhl for sharing data of forensic interest. Finally, students and fellow researchers at *Ca' Foscari University of Venice* and the *University of Lausanne* have provided the authors with exciting and encouraging environments without which much of the writing of this book would not have been possible.

Venice, Italy   Silvia Bozza
Lausanne-Dorigny, Switzerland   Franco Taroni
Lausanne-Dorigny, Switzerland   Alex Biedermann
August 2022

Contents

Chapter 1
Introduction to the Bayes Factor and Decision Analysis

1.1 Introduction

The assessment of the value of scientific evidence involves subtle forensic, statistical, and computational aspects that can represent an obstacle in practical applications. The purpose of this book is to provide theory, examples, and elements of R code to illustrate a variety of topics pertaining to value of evidence assessments using Bayes factors in a decision-theoretic perspective.

The structure of this book is as follows. This chapter starts by presenting an overview of the role of statistics in forensic science, with an emphasis on the Bayesian perspective and the role of the Bayes factor for logical inference and decision. Next, the chapter addresses three general topics that forensic scientists commonly encounter: model choice, evaluation, and investigation. For each of these themes, Bayes factors will be developed and discussed using practical examples. Particular attention will be devoted to the distinction between feature- and score-based Bayes factors, typically used in evaluative settings. This chapter also provides theoretical background analysts might need during data analysis, including elements of forensic interpretation, computational methods, decision theory, prior elicitation, and sensitivity analysis.

Chapter 2 addresses the problem of discrimination between competing propositions regarding target features of a population of interest (i.e., parameters). Examples include applications involving counting processes and propositions referring to the proportion of items of forensic interest (e.g., items with illegal content) or an unknown quantity. Attention will be drawn to background elements that may affect counting processes or continuous measurements and a decisional approach to this problem.

Chapter 3 addresses the problem of evaluation of scientific evidence in the form of discrete, continuous, and continuous multivariate data. The latter may present a complex dependence structure that will be handled by means of multilevel models.

S. Bozza et al., *Bayes Factors for Forensic Decision Analyses with R*, Springer Texts in Statistics, https://doi.org/10.1007/978-3-031-09839-0_1

Chapter 4 focuses on the problem of investigation, using examples involving either univariate or multivariate data.

For each topic covered in the book, examples will be accompanied with R code, allowing readers to reproduce computations and adapt sample code to their own problems. The end of each chapter presents an outline of the principal R functions used throughout the respective chapters. While some functions can be easily reproduced, others are more elaborate and copying their R code would be tedious. These functions are available, as well as datasets, as supplementary materials on the book's website (on http://link.springer.com/).

1.2   Statistics in Forensic Science

Forensic science uses scientific principles and technical methods to help with the use of evidence in legal proceedings of criminal, civil, or administrative nature. To assist members of the judiciary in their inquiries regarding the existence or past occurrence of events of legal interest, forensic scientists examine recovered traces, objects, and materials related to persons of interest. This may involve, for example, the analysis of the nature of body fluids and various other items such as textile fibers, glass and paint fragments, handwriting, digital device data, as well as the classification of such items and data into various categories.

More generally, forensic science takes a major interest in both investigative proceedings and evaluative processes at trial. This involves the examination of persons and objects, as well as the vestiges of actions. Forensic scientists also help with reconstructing past events. Thus, incomplete knowledge and, hence, uncertainty are key challenges that all participants in the legal process must deal with. The standard approach to cope with uncertainty is the structured collection and sound use of data. Typically, data result from the analysis and comparative examination of evidential material (i.e., biological traces, toxic substances, documents, crime scene findings, imaging data, etc.), followed by an assessment of the probative value of scientific results within the context of the event under investigation and in the light of the task-relevant information.

However, despite its potential to support legal evidence and proof processes, forensic science has also been found to be a contributing factor to miscarriages of justice (Cole, 2014). Furthermore, over the last decade, reviews by expert panels have exposed several areas of forensic science practice as insufficiently reliable (e.g., PCAST, 2016), and courts across many jurisdictions have insisted on the need to probe and demonstrate the empirical foundations of forensic science disciplines.

Scientists currently address these challenges by directing research not only toward more studies involving experiments under controlled conditions but also by developing formal frameworks for value of evidence assessment that can cope with scientific evidence independent of its nature and type. Central to this development is a convergence to the Bayesian perspective, which is well suited to help forensic scientists assess the probative value of observations that, typically, do not arise under only one given hypothesis or proposition.[1] Bayesian thinking can cope with situations in which one holds varying degrees of belief about competing hypotheses and one considers that those hypotheses may differ in their capacity to account for one's observations and findings. As noted by Cornfield (1967, p. 34),

> Bayes' theorem is important because it provides an explication for this process of consistent choice between hypotheses on the basis of observations and for quantitative characterization of their respective uncertainties.

In forensic science, the *Bayes factor* (BF)—a central element in Bayesian analysis—has come to play an extremely important role. It represents a key statistic for assessing the value of scientific findings and is, thereby, widely covered in forensic literature (e.g., Aitken et al., 2021; Buckleton et al., 2016). It allows scientists to assess case-related observations or measurements in the light of competing propositions presented by parties at trial. In essence, the Bayes factor is a concept that provides a measure of the degree to which a scientific finding is capable to discriminate between the competing propositions of interest.

The choice of the Bayes factor to assess the value of outcomes of laboratory examinations and analyses results from the requirement to comply with several practical precepts of coherent thinking and decision-making. The desirable properties that the Bayes factor accounts for are balance, transparency, robustness, and logic. In addition, it is a flexible measure, acknowledged throughout forensic science, law, and statistics, because it can deal with any type of evidence (e.g., Evett, 1996; Jackson, 2000; Robertson & Vignaux, 1993; Robertson et al., 2016; Good, 1950; Kass & Raftery, 1995; Lindley, 1977; Taroni et al., 2010).

In forensic science, the Bayes factor is more commonly called *likelihood ratio*, even if this may create confusion because the two terms represent two distinct concepts, and the Bayes factor does not always simplify to a likelihood ratio. This will be explained later in Sect. 1.4. Generally, the use of the Bayes factor is now well established in both theory and practice, though some branches of forensic science are more advanced in Bayes factor analyses than others. A general overview is presented by the Royal Statistical Society's Section Committee on Statistics and Law (e.g., Aitken et al., 2010) in a series of practitioner guides for judges, forensic scientists, and expert witnesses.

While the Bayes factor represents a coherent metric for value of evidence

---

[1] The term hypothesis (or proposition) is interpreted here as an assertion or a statement that such and such is the case (e.g., an outcome or a state of nature of the kind "the questioned document has been printed with printer 1" or "the recovered item is from the same source as the control item") and also as a description of a decision. Propositions are, therefore, statements that are either true or false and that can be affirmed or denied. An important basis for much of the argument developed in this book is the assumption that personal degrees of belief can be assigned to propositions or hypotheses. Throughout this book, hypothesis and proposition are treated as synonyms.

assessment[2] in evaluative reporting[3] (i.e., when a person of interest is available for comparison purposes), it is important to mention that it can also be used in investigative contexts. A case is investigative when there is no person or object available for comparison, and examinations concentrate primarily on helping to draw inferences about general features (e.g., sex, right-/left-handedness, etc.) related to the source of a recovered stain, mark, or trace. More generally, the Bayes factor can be used for two main purposes in forensic science:

- The first purpose is to assign a value to the result of a comparison between an item of unknown source and an item from a known source. This refers to the *evaluative* mode in which forensic scientists operate. Evaluating a scientific finding thus means that the scientist provides an expression of the value of the observation in support—which may be positive, negative, or neutral—of a proposition of interest in legal proceedings, compared to a relevant alternative proposition.
- The second purpose is to provide information in investigative proceedings. Here, scientists operate in what is called *investigative* mode. They try to help answer questions such as "what happened?" and "what (material) is this?" (Jackson et al., 2006). The scientist is said to be "event focused" and uses the findings to generate hypotheses and suggestions for explanations of observations, in order to give guidance to investigators or litigants.

To illustrate these concepts, imagine a case involving a questioned document and handwriting. In cases of anonymous letter-writing, it regularly occurs that, at least initially, no suspected writer is available. In such a case, there will be no possibility for jointly evaluating characteristics observed on a questioned document and features on reference (known or control) material from a person of interest, as would be the case in an evaluative context. However, this does not mean that measurements made only on the questioned document, without comparison to reference material, could not be informative for investigative purposes. For example, features extracted from the handwriting of unknown source may be evaluated with respect to more general propositions such as "the questioned document (e.g., a ransom note) has been written by a man (woman)" or "the questioned document has been written by a right- (left)-handed person." Helping to discriminate between such propositions contributes to reducing the pool of potential writers in an investigation.

As a metric to assess the value of findings in a forensic context, the Bayes factor allows practitioners to offer a quantitative expression that they can convey in a more general reasoning framework that conforms to the logic of Bayesian thinking. From the scientist's point of view, the contribution to inference is perfectly symmetric. That is, the findings may support either of the two competing propositions, with

---

[2] A list of necessary logical conditions to guarantee coherence is presented and discussed in Taroni et al. (2021a).

[3] On the difference between evaluative and other types of reporting, such as technical and intelligence reporting, see ENFSI Guideline for Evaluative Reporting in Forensic Science (Willis et al., 2015) §1.1.

respect to the relevant alternative proposition. This strengthens the scientist's role as balanced expert in the legal process.

1.3 Bayesian Thinking and the Value of Evidence

Bayesian philosophy is named after Reverend Thomas Bayes and is based on an interpretation of probability as personal degree of belief (de Finetti, 1989). In Bayesian theory, all uncertainties in a problem must necessarily be described by probabilities. Probability is intended as one's conditional measure of uncertainty associated with the evidence, the available information, and all the underlying assumptions. In this book, we will use the term evidence in the general sense of a given piece of information or data. This includes, but is not restricted to, the idea of evidence used in legal proceedings. The term evidence is used here in a broad sense as synonym for other terms such as "finding" or "outcome." According to Good (1988), evidence may be defined as data that makes one alter one's beliefs about how the world is working. The word finding, in turn, is used in this book to designate the result of a forensic examination or analysis. Findings are measurements in a quantitative form, discrete or continuous. Examples for discrete quantitative results are counts of glass fragments or gunshot residues. Examples for continuous results are measurements of physical quantities such as length, weight, refractive index, and summaries of complex comparisons in the form of similarity scores. For a formal definition of the term findings, see also the ENFSI Guideline for Evaluative Reporting in Forensic Science (Willis et al., 2015).

Starting from prior probabilities, representing subjective degrees of belief about propositions of interest, the Bayesian paradigm allows one to rationally revise such beliefs and compute posterior probabilities, draw inferences about propositions, and make decisions (Sprenger, 2016). For example, when new information becomes available, it may be necessary to assess how this information ought to affect propositions regarding the involvement of a person of interest in particular alleged activities. Likewise, physicians need to structure their thought processes when performing medical diagnosis. In general, the question is how to update one's personal beliefs regarding uncertain events when one receives new information.

Suppose that the events $H_1, \ldots, H_n$ form a partition, and denote by $\Pr(H_i \mid I)$ the probability that is associated with H_i, $i = 1, \ldots, n$, given relevant background information I. This probability is called a *prior probability*. Furthermore, consider an event or quantity E, whose probability can be expressed by means of the *law of total probability* as

$$\Pr(E \mid I) = \sum_j \Pr(E \mid H_j, I) \Pr(H_j \mid I). \tag{1.1}$$

The ENFSI Guideline for Evaluative Reporting in Forensic Science (Willis et al., 2015, at p. 21) regards conditioning information as the essential ingredient of probability assignment, since all probabilities are conditional. In forensic evaluation, it is important not to focus on all possible information, but only on the information that is relevant to the forensic task at hand. Disciplined forensic reporting requires scientists to make clear their perception of the conditioning information at the time they conduct their evaluation. Conditioning information is sometimes known as the framework of circumstances (or background information). Much of the non-scientific information will not have a bearing on the value of scientific findings, but it is essential to recognize those aspects that do. Examples of relevant information may include the ethnic origin of the perpetrator (but not that of the suspect) and the nature of garments and surfaces involved in alleged transfer events. More generally, conditioning information may also include data and domain knowledge that the expert uses to assign probabilities. The conditioning on (task-) relevant information I is important because it clarifies that probability assignments are personal and depend on the knowledge of the person conducting the evaluation.

Bayes rule (or theorem) is a straightforward application of the conditionalization principle and the partition formula (1.1). It allows one to compute the so-called *posterior probability* $Pr(H_i \mid E, I)$ as

$$Pr(H_i \mid E, I) = \frac{Pr(E \mid H_i, I) Pr(H_i \mid I)}{Pr(E \mid I)} = \frac{Pr(E \mid H_i, I) Pr(H_i \mid I)}{\sum_j Pr(E \mid H_j, I) Pr(H_j \mid I)},$$

which emphasizes that certain knowledge of E modifies the probability of H_i.[4] Note that *prior* and *posterior* probabilities are only relative to the new finding E. The posterior probability will become again a prior probability when additional findings become available. Lindley (2000, p. 301) expressed this as follows: "Today's posterior is tomorrow's prior." Bayesian statistics is the sequential application of Bayes rule to all situations that involve observed and missing data, unknown quantities (e.g., events, propositions, population parameters), or unobserved data (e.g., future observations).

Participants in the legal process are typically concerned with the problem of comparing competing propositions about a contested event. A typical example for trace evidence is "the recovered glass fragments come from the broken window" versus "the recovered glass fragments come from an unknown source." When measurements on various items (i.e., glass fragments) are available, it may be necessary to quantitatively evaluate these findings with respect to selected propositions of interest. According to Bayesian methodology developed by Jeffreys (1961), this involves the introduction of a statistical model to describe the probability of the available measurements according to different hypotheses (propositions or models). The posterior probability of each hypothesis is then computed via a direct application of Bayes theorem. Following Jeffreys' criterion for comparing hypotheses, a hypothesis is accepted or rejected on the basis of its posterior

---

[4] See Taroni et al. (2020) for a discussion on the generalization of Bayes rule (i.e. Jeffrey's conditionalization) when one is faced to uncertain evidence.

probability being greater or smaller than that of the alternative proposition. Note that the acceptance or rejection of a proposition is not meant as an assertion of its truth or falsity, only that its probability is greater or smaller than that of the respective alternative proposition (Press, 2003).

The primary element in Bayesian methodology for comparing propositions is the Bayes factor (BF for short). It provides a numerical representation of the impact of findings on propositions of interest. In other words, the Bayes factor quantifies the degree to which observed measurements discriminate between competing propositions. The Bayes factor is the ingredient by which the prior odds in favor of a proposition are multiplied in virtue of the knowledge of the findings (Good, 1958):

$$\text{Posterior odds} = \text{BF} \times \text{Prior odds.}$$

Broadly speaking, prior and posterior odds are the ratios of probabilities of the hypotheses of interest before and after acquiring new findings, respectively. The value of experimental outcomes is measured by how much *more* probable they make one hypothesis relative to the respective alternative hypothesis, compared to the situation *before* considering the experimental findings.

A formal definition of the Bayes factor is given in Sect. 1.4, along with a discussion about its interpretation as measure of the value of the evidence. Practical examples in Sects. 1.5 and 1.6 and further developments in Chaps. 3 and 4 will illustrate the use of the Bayes factor for evaluative and investigative purposes.

1.4 Bayes Factor for Model Choice

Consider an unknown quantity X, referring to a quantity or measurement of interest such as the number of ecstasy pills in a sample drawn from a large seizure of pills, the elemental chemical composition of glass fragments, or a feature (e.g., the length) of a handwritten character. Furthermore, suppose that $f(x \mid \theta)$ is a suitable *probability model*[5] for X, where the unknown parameter[6] θ belongs to the parameter space Θ. Suppose also that the parameter space consists of two non-overlapping sets Θ_1 and Θ_2 such that $\Theta = \Theta_1 \cup \Theta_2$. A question that may be of interest is whether the parameter θ belongs to Θ_1, or to Θ_2, that is to compare the hypothesis

$$H_1 : \theta \in \Theta_1,$$

against the alternative hypothesis

---

[5] A probability model is understood here as a characterization of the distribution of measurements.

[6] A parameter is taken here as a characteristic of the distribution of all members (e.g., individuals or objects) of a population of interest.

$$H_2 : \theta \in \Theta_2.$$

Note that H_1 is usually called the null hypothesis. Under a classical (frequentist) approach, the distinction between null and alternative hypotheses is very important. Users must be aware that when performing significance testing, competing hypotheses are not equivalent and there is, in fact, an asymmetry associated with them. One collects data (or evidence) against the null hypothesis before it is rejected, but the acceptance of the null hypothesis is not an assertion about its truthfulness. It merely means that there is little evidence against it. As will be shown, under the Bayesian paradigm, this does not represent an issue.

A hypothesis H_i is called *simple* if there is only one possible value for θ, say $\Theta_i = \{\theta_i\}$. A hypothesis is called *composite* (see, e.g., Example 1.1) if there is more than one possible value.

Let $\pi_1 = \Pr(H_1) = \Pr(\theta \in \Theta_1)$ and $\pi_2 = \Pr(H_2) = \Pr(\theta \in \Theta_2)$ denote the prior probabilities for the competing composite hypotheses H_1 and H_2. Note that, for the sake of simplicity, the letter I denoting background information is omitted here. The ratio of the prior probabilities π_1/π_2 is called the *prior odds* of H_1 to H_2. The prior odds indicate whether hypothesis H_1 is more or less probable than hypothesis H_2 (prior odds being greater or smaller than 1) or whether the hypotheses are (almost) equally probable, i.e., the prior odds are (close) to 1.[7] Suppose observational data x are available that do not provide conclusive evidence[8] about the propositions of interest but will allow one to update prior beliefs using Bayes theorem. Let us denote by $f_{H_i}(x)$ the *marginal probability* of the data under proposition H_i, that is,

$$f_{H_i}(x) = \int_{\Theta_i} f(x \mid \theta)\pi_{H_i}(\theta)d\theta, \qquad (1.2)$$

where $\pi_{H_i}(\theta)$ denotes the prior probability density of θ for $\theta \in \Theta_i$. The marginal probability is also called the *predictive probability*, which is the probability to observe the actual data before any data become available. Kass and Raftery (1995) refer to it as the *marginal likelihood*: the probability of the observations averaged

---

[7] The ratio of the probabilities of two mutually exclusive and collectively exhaustive events is called *odds* in favor of the event whose probability is in the numerator of the ratio. Note that hypotheses are not necessarily exhaustive: the word odds is sometimes used loosely in reference to the ratio of the probabilities of mutually exclusive propositions whose probabilities do not add to 1 (Taroni et al., 2010).

[8] The problem of imperfect evidence is well illustrated by Robertson and Vignaux (1995, at p.12):

An ideal piece of evidence would be something that always occurs when what we are trying to prove is true and never occurs otherwise. If we are trying to demonstrate the truth of an hypothesis or assertion we would like to find as evidence something which always occurs when the hypothesis is true and never occurs when the hypothesis is not true. In real life, evidence this good is almost impossible to find.

across the prior distribution over the parameter space Θ. Note that the parameter space Θ can be either continuous or discrete. In the latter case, the integral in (1.2) must be replaced by a sum, and the marginal probability of the evidence (i.e., data x) becomes

$$f_{H_i}(x) = \sum_{\theta \in \Theta_i} f(x \mid \theta) \Pr(\theta \mid H_i).$$

The Bayes factor for comparing H_1 and H_2 is defined as the ratio of the marginal probabilities $f_{H_i}(x)$ under the competing hypotheses, that is,

$$BF = \frac{f_{H_1}(x)}{f_{H_2}(x)}. \tag{1.3}$$

Let $\alpha_1 = \Pr(H_1 \mid x) = \Pr(\theta \in \Theta_1 \mid x)$ and $\alpha_2 = \Pr(H_2 \mid x) = \Pr(\theta \in \Theta_2 \mid x)$ denote the posterior probabilities for the competing hypotheses. The ratio of the posterior probabilities α_1/α_2 is called the *posterior odds* of H_1 to H_2. Recalling the odds form of Bayes theorem, one can express the Bayes factor for comparing hypothesis H_1 against hypothesis H_2 as the factor by which the prior odds of H_1 to H_2 are multiplied in virtue of the knowledge of the data to obtain the posterior odds, that is,

$$\alpha_1/\alpha_2 = BF \times \pi_1/\pi_2.$$

The Bayes factor measures the change produced by the new information (or, data) in the odds when going from the prior to the posterior distributions in favor of one proposition as opposed to a given alternative. For this reason, it is not uncommon to find the BF defined as the ratio of the posterior odds in favor of H_1 to the prior odds in favor of H_1, that is,

$$BF = \frac{\alpha_1/\alpha_2}{\pi_1/\pi_2}. \tag{1.4}$$

One of the attractive features of using a Bayes factor to quantify the value of the acquired information is that it does not depend on prior probabilities of competing hypotheses. However, this bears potential for misunderstandings. The Bayes factor is sometimes interpreted as, for example, the odds provided by the data alone, for H_1 to H_2: this is conceptually incorrect. Though cases may be found where the Bayes factor can be expressed as a ratio of likelihoods[9] and correctly be interpreted

---

[9] While probabilistic modeling provides the probability $f(x \mid \theta)$ of any hypothetical data x before any observation is made, conditional on θ, statistical methods allow one to draw conclusions about θ given the collected observations x. This difference in focus is expressed by the *likelihood function*, written $l(\theta \mid x)$, where the probability distribution $f(x \mid \theta)$ is written as a function of θ conditional on the observations x, i.e., $f(x \mid \theta) = l(\theta \mid x)$.

as the "summary of the evidence provided by the data in favor of one scientific theory (...) as opposed to another" (Kass & Raftery, 1995, at p. 777), this does not hold in general. The Bayes factor will generally depend on prior assumptions. It is necessary, thus, to clarify the meaning of "prior assumptions" because confusion may arise between, on the one hand, the notion of prior probability about model parameters ($\theta \in \Theta_i$) and, on the other hand, prior probabilities of propositions (H_i).

To clarify this distinction, consider the comparison of a simple hypothesis H_1 : $\theta = \theta_1$ against a simple alternative hypothesis $H_2 : \theta = \theta_2$. The prior probabilities of these hypotheses are expressed as $\pi_1 = \Pr(\theta = \theta_1)$ and $\pi_2 = \Pr(\theta = \theta_2)$. The posterior probabilities α_i in the light of prior probabilities π_i ($i = 1, 2$) and observed data x can be easily computed by means of a direct application of Bayes theorem:

$$\alpha_i = \Pr(H_i \mid x) = \Pr(\theta = \theta_i \mid x) = \frac{f(x \mid \theta_i)\pi_i}{\sum_{j=1,2} f(x \mid \theta_j)\pi_j}. \tag{1.5}$$

The ratio of the posterior probabilities α_1/α_2 obtained from computing (1.5) for $i = 1, 2$ simplifies to the product of the likelihood ratio times the ratio of the prior probabilities, that is,

$$\frac{\alpha_1}{\alpha_2} = \frac{f(x \mid \theta_1)}{f(x \mid \theta_2)} \times \frac{\pi_1}{\pi_2}.$$

Recalling (1.4), it is readily seen that the Bayes factor in this simple case is the likelihood ratio of H_1 to H_2,

$$\mathrm{BF} = \frac{f(x \mid \theta_1)}{f(x \mid \theta_2)} \times \frac{\pi_1}{\pi_2} \times \frac{\pi_2}{\pi_1} = \frac{f(x \mid \theta_1)}{f(x \mid \theta_2)}, \tag{1.6}$$

and it is correct then to interpret this as "the odds provided by the data alone for H_1 to H_2."

However, the comparison of simple versus simple hypotheses is a particular case among many others. Practitioners may face the more general situation where at least one of the hypotheses is composite, that is, the parameter of interest may take one of a range of different values (e.g., $\Theta_i = \{\theta_1, \ldots, \theta_k\}$), or infinitely many, as is the case when θ is continuous. In the case of composite hypotheses, the prior probabilities π_i for $i = 1, 2$ will take the following form:

$$\pi_i = \Pr(\theta \in \Theta_i) = \begin{cases} \sum_{\theta \in \Theta_i} \Pr(\theta) & \text{for } \theta \text{ discrete} \\ \int_{\Theta_i} \pi(\theta)d\theta & \text{for } \theta \text{ continuous,} \end{cases} \tag{1.7}$$

where $\pi(\theta)$ is the prior probability density for $\theta \in \Theta$. The posterior probabilities α_i are therefore computed as

$$\alpha_i = \Pr(\theta \in \Theta_i \mid x) = \begin{cases} \dfrac{\sum_{\theta \in \Theta_i} f(x|\theta) \Pr(\theta)}{\sum_{\theta \in \Theta} f(x|\theta) \Pr(\theta)} & \text{for } \theta \text{ discrete} \\[4ex] \dfrac{\int_{\Theta_i} f(x|\theta) \pi(\theta) d\theta}{\int_{\Theta} f(x|\theta) \pi(\theta) d\theta} & \text{for } \theta \text{ continuous,} \end{cases} \tag{1.8}$$

and the posterior odds will be

$$\frac{\alpha_1}{\alpha_2} = \begin{cases} \dfrac{\sum_{\theta \in \Theta_1} f(x|\theta) \Pr(\theta)}{\sum_{\theta \in \Theta_2} f(x|\theta) \Pr(\theta)} & \text{for } \theta \text{ discrete} \\[4ex] \dfrac{\int_{\Theta_1} f(x|\theta) \pi(\theta) d\theta}{\int_{\Theta_2} f(x|\theta) \pi(\theta) d\theta} & \text{for } \theta \text{ continuous.} \end{cases} \tag{1.9}$$

Following (1.4), the Bayes factor can be reconstructed as follows:

$$\text{BF} = \begin{cases} \dfrac{\sum_{\theta \in \Theta_1} f(x|\theta) \Pr(\theta)}{\sum_{\theta \in \Theta_2} f(x|\theta) \Pr(\theta)} \Big/ \dfrac{\pi_1}{\pi_2} & \text{for } \theta \text{ discrete} \\[4ex] \dfrac{\int_{\Theta_1} f(x|\theta) \pi(\theta) d\theta}{\int_{\Theta_2} f(x|\theta) \pi(\theta) d\theta} \Big/ \dfrac{\pi_1}{\pi_2} & \text{for } \theta \text{ continuous,} \end{cases} \tag{1.10}$$

where the π_i are computed as in (1.7). It is seen that the Bayes factor can no longer be expressed as a likelihood ratio as in the case of comparing simple versus simple hypotheses. We will show this for the case where θ is continuous.

Start with the prior probability density $\pi(\theta)$ on Θ, and divide it by the probability π_i of the hypothesis H_i to obtain the restriction of the prior probability density $\pi(\theta)$ on Θ_i, that is,

$$\pi_{H_i}(\theta) = \frac{\pi(\theta)}{\pi_i} \quad \text{for } \theta \in \Theta_i.$$

The probability density $\pi_{H_i}(\theta)$ simply describes how the prior probability spreads over the hypothesis H_i. The prior probability density $\pi(\theta)$ can thus be rewritten in the following form:

$$\pi(\theta) = \begin{cases} \pi_1 \pi_{H_1}(\theta) \text{ for } \theta \in \Theta_1, \\[2ex] \pi_2 \pi_{H_2}(\theta) \text{ for } \theta \in \Theta_2. \end{cases}$$

Therefore, the posterior odds in (1.9) for the continuous case can be rewritten as

$$\frac{\alpha_1}{\alpha_2} = \frac{\pi_1 \int_{\Theta_1} f(x \mid \theta) \pi_{H_1}(\theta) d\theta}{\pi_2 \int_{\Theta_2} f(x \mid \theta) \pi_{H_2}(\theta) d\theta}. \tag{1.11}$$

Recalling (1.4), the Bayes factor in (1.10) will take the form of integrated likelihoods under the hypotheses of interest, that is,

$$\text{BF} = \frac{\int_{\Theta_1} f(x \mid \theta) \pi_{H_1}(\theta) d\theta}{\int_{\Theta_2} f(x \mid \theta) \pi_{H_2}(\theta) d\theta}. \tag{1.12}$$

The reader can verify that the two expressions in (1.3) and (1.12) are equivalent. Prior evaluations enter the Bayes factor through the weights $\pi_{H_1}(\theta)$ and $\pi_{H_2}(\theta)$. The Bayes factor depends on how the prior mass is spread over the two hypotheses (Berger, 1985). It is also worth noting that whenever hypotheses are unidirectional (e.g., when comparing $H_1 : \theta \leq \theta_0$ against $H_2 : \theta > \theta_0$), the choice of a prior probability density $\pi(\theta)$ over $\Theta = \Theta_1 \cup \Theta_2$ (with $\Theta_1 = [0, \theta_0]$ and $\Theta_1 = (\theta_0, 1]$) is equivalent to the expression of a prior probability for the competing hypotheses. Conversely, whenever hypotheses are bidirectional (e.g., when comparing $H_1 : \theta = \theta_0$ against $H_2 : \theta \neq \theta_0$), one cannot choose a prior probability density $\pi(\theta)$ over the entire parameter space Θ, as this would amount to place a probability equal to 0 to the hypothesis $H_1 : \theta = \theta_0$. The prior probability distribution over θ must, in this case, be a mixture of a discrete component that assigns a positive mass $\pi_1 = \Pr(\theta = \theta_0)$ to H_1 and a continuous component that spreads the remaining mass $\pi_2 = 1 - \pi_1$ over Θ_2 according to the probability density $\pi_{H_2}(\theta)$. The posterior probability α_1 can then be computed as in (1.8), where $\Theta_1 = \theta_0$,

$$\alpha_1 = \Pr(H_1 \mid x) = \frac{\pi_1 f(x \mid \theta_0)}{\pi_1 f(x \mid \theta_0) + \pi_2 \int_{\Theta_2} f(x \mid \theta) \pi_{H_2}(\theta) d\theta}. \tag{1.13}$$

Analogously, the posterior probability α_2 may be computed, and the Bayes factor is

$$\text{BF} = \frac{f(x \mid \theta_0)}{\int_{\Theta_2} f(x \mid \theta) \pi_{H_2}(\theta) d\theta}. \tag{1.14}$$

It can be observed that the Bayes factor in (1.14) does not depend on the prior probabilities of competing hypotheses which can vary considerably among recipients of expert information. Any such recipient can, starting from their own probabilities, use the Bayes factor to obtain posterior probabilities in a straightforward manner. Consider, for the sake of illustration, the posterior probability of hypothesis H_1 in (1.13). A simple manipulation allows one to obtain

$$\alpha_1 = \left[1 + \frac{\pi_2}{\pi_1} \frac{1}{\text{BF}} \right]^{-1} = \frac{\text{BF}}{\text{BF} + \pi_2/\pi_1}.$$

In summary, the Bayes factor thus measures the change in the odds in favor of one hypothesis, as compared to a given alternative hypothesis, when going from the prior to the posterior distribution. This means that a Bayes factor larger than 1 indicates that the data support hypothesis H_1 compared to H_2. However, the Bayes factor does not indicate whether H_1 is more *probable* than the opposing hypothesis H_2, the BF only makes it more probable than it was *before observing* the data (Lavine & Schervish, 1999).

Example 1.1 (Alcohol Concentration in Blood) A person is stopped because of suspicion of driving under the influence of alcohol. Blood taken from that person is submitted to a forensic laboratory to investigate whether the quantity of alcohol in blood θ is greater than a legal threshold of, say, 0.5 g/kg. Thus, the hypotheses of interest can be defined as $H_1 : \theta > 0.5$ versus $H_2 : \theta \leq 0.5$. Suppose that a prior probability density $\pi(\theta)$ is given for θ and that the prior probabilities of H_1 and H_2 in (1.7) are $\pi_1 = 0.05$ and $\pi_2 = 0.95$, corresponding to prior odds approximately equal to 0.0526. These values suggest that, based on the circumstances, and before considering results of blood analyses, the hypothesis H_1 is believed to be much less probable than the alternative hypothesis. Suppose next that the posterior probabilities, after taking into account laboratory measurements, are computed as in (1.8). The results are $\alpha_1 = 0.24$ and $\alpha_2 = 0.76$. Thus, the posterior odds are approximately equal to 0.3158. The ratio of the posterior odds by the prior odds leads to a BF equal to 6. This result represents limited evidence in support of the hypothesis that the alcohol level in blood is greater than the legal threshold, compared to the alternative hypothesis. Still, the posterior probability of hypothesis H_1 is low: the BF only renders the hypothesis H_1 slightly more probable than it was before observing the measurements made in the laboratory. This example will be further developed in Chap. 2.

1.5   Bayes Factor in the Evaluative Setting

Consider the general situation where evidentiary material is collected and control items from a person or object of interest are available for comparative purposes. The following measurements of a particular characteristic are available: measurements y on a questioned item (e.g., a glass fragment found on the clothing of a person of interest) and measurements x on a control item (e.g., fragments from a broken window). In this evaluative setting, so-called source level propositions[10] could be defined as follows:

---

[10] The notion of *source level* refers to a given level in a hierarchy of hypotheses. This view considers a classification (i.e., hierarchy) of propositions into three main categories or levels, called the source level, activity level, and crime level. See Cook et al. (1998) for a discussion. Note that source level propositions for the example of glass fragments are chosen here as a formative example and for illustrative purposes. As a type of transfer evidence, glass fragments should be evaluated using activity level propositions (Willis et al., 2015).

H_1:   The recovered (i.e., questioned) item comes from the same source as the control item.

H_2:   The recovered (i.e., questioned) item comes from an unknown source (i.e., different from the control item).

This setting is called evaluative because it involves the comparison between control and recovered items and the use of the results of this comparison for discriminating between the competing propositions. Models for comparison can either be *feature-based* or *score-based*. Feature-based models (Sect. 1.5.1) focus on the probability of measurements made directly on evidentiary and reference items. Conversely, score-based models (Sect. 1.5.2) focus on the probability of observing a pairwise similarity (or distance), i.e., score, between compared materials.

1.5.1   Feature-Based Models

If one assumes that y and x are realizations of random variables Y and X with a given probability distribution $f(\cdot)$, the Bayes factor is

$$\mathrm{BF} = \frac{f(y, x \mid H_1, I)}{f(y, x \mid H_2, I)}, \tag{1.15}$$

where I represents the available background information. Application of the rules of conditional probability allows one to rewrite the Bayes factor as follows:

$$\mathrm{BF} = \frac{f(y \mid x, H_1, I)}{f(y \mid x, H_2, I)} \times \frac{f(x \mid H_1, I)}{f(x \mid H_2, I)}.$$

This expression can be further simplified by considering the fact that (i) the distribution of measurements x on the control item does not depend on whether H_1 or H_2 is true (and hence $f(x \mid H_1, I) = f(x \mid H_2, I)$ holds) and (ii) the distribution of the measurement y on the questioned item does not depend on the measurement x on the control item if H_2 is true,[11] so that $f(y \mid x, H_2, I) = f(y \mid H_2, I)$. The Bayes factor can therefore be written as

$$\mathrm{BF} = \frac{f(y \mid x, H_1, I)}{f(y \mid H_2, I)}. \tag{1.16}$$

---

[11] Note that this assumption of independence is not always valid, e.g., with DNA evidence (Balding & Nichols, 1994; Aitken et al., 2021). A further example is the case of questioned signatures. Under the proposition that a signature has been forged and therefore is not authentic, one should take into account that a forger will attempt to reproduce the features of a target signature. Thus, recovered and control measurements cannot be considered independent (Linden et al., 2021); see Sect. 3.4.3.

The numerator is the probability of observing the measurements on the recovered item under the assumption that it comes from the known source, given the information I and knowledge of x, the features of the known source. The denominator is the probability of observing the measurements y on the recovered item, assuming that it comes from an unknown source, usually selected in some aleatory way from a relevant population,[12] and assuming again the relevant information I. Note that, for the sake of simplicity, the conditioning information I will be omitted in the arguments hereafter.

For many types of forensic evidence, it can be reasonable to assume a parametric model $\{f(\cdot \mid \theta), \theta \in \Theta\}$. In this way, the probability distribution characterizing the available data is of a known form, with the only unknown element being the parameter θ, which may vary between sources. Consider, for example, the probability distribution $f(\cdot \mid \theta)$ with unknown parameter $\theta = \theta_y$ for the measurements y on the recovered item and the same probability distribution with unknown parameter $\theta = \theta_x$ for the measurements x on the control item. In practice, the parameter θ is unknown, and a prior probability distribution $\pi(\theta \mid H_i)$, representing personal beliefs about θ under each hypothesis H_i, is introduced. The marginal distribution $f(y \mid x, H_1)$ in the numerator of (1.16) may be rewritten as follows:

$$f(y \mid x, H_1) = \int f(y \mid \theta)\pi(\theta \mid x, H_1)d\theta$$

$$= \int f(y \mid \theta)f(x \mid \theta)\pi(\theta \mid H_1)d\theta / f(x \mid H_1), \qquad (1.17)$$

where the posterior density $\pi(\theta \mid x, H_1)$ in the first line is rewritten in extended form using Bayes theorem. The distribution $f(y \mid x, H_1)$ is also called a *posterior predictive* distribution.[13]

The marginal distribution $f(y \mid H_2)$ in the denominator of (1.16) can be rewritten as follows:

$$f(y \mid H_2) = \int f(y \mid \theta)\pi(\theta \mid H_2)d\theta. \qquad (1.18)$$

This is also called a predictive distribution.

---

[12] Note that rules of conditional probability do not specify on which variable we should condition. Champod et al. (2004) suggest that we should condition on the item with greater information content. Therefore we usually condition on the control item (e.g., in the case of DNA, traces can be degraded or of small quantity, while a complete profile can usually be obtained for a person of interest). For further comments, see also Aitken et al. (2021, pp. 619–627).

[13] For a discussion of posterior predictive distributions in forensic science contexts, see, e.g., Biedermann et al. (2015).

Example 1.2 (Toner on Printed Documents) Suppose experimental findings are available in the form of measurements of magnetism of toner on printed documents of known origin (x) and questioned origin (y) for which a Normal distribution is considered suitable. Thus, $X \sim N(\theta_x, \sigma^2)$ and $Y \sim N(\theta_y, \sigma^2)$, where the variance σ^2 of both distributions is assumed known and equal (Biedermann et al., 2016a). A Normal distribution with mean μ and variance τ^2 is taken to model our prior uncertainty about the means θ_x and θ_y, that is, $\theta \sim N(\mu, \tau^2)$ for $\theta = \{\theta_x, \theta_y\}$. The integrals in (1.17) and (1.18) have an analytical solution, and the marginals can be obtained in closed form. See Aitken et al. (2021, pp. 815–817) for more details.

Here, H_1 and H_2 denote the propositions according to which the items of toner come from, respectively, the same and different printing machines. Consider, first, the numerator of the BF in (1.17), where the posterior density $\pi(\theta \mid x, H_1)$ is still a Normal distribution with mean μ_x and variance τ_x^2, computed according to well-known updating rules (see, e.g., Lee, 2012),

$$\mu_x = \frac{\sigma^2}{\sigma^2 + \tau^2}\mu + \frac{\tau^2}{\sigma^2 + \tau^2}x \tag{1.19}$$

and

$$\tau_x^2 = \frac{\sigma^2 \tau^2}{\sigma^2 + \tau^2}. \tag{1.20}$$

The posterior mean, μ_x, is a weighted average of the prior mean μ and the observation x. The weights are given by the population variance σ^2 and the variance τ^2 of the prior probability distribution, respectively, such that the component (observation or prior mean) which has the smaller variance has the greater contribution to the posterior mean. This result can be generalized to consider the distribution of the mean of a set of n observations $x_1, \ldots, x_n$ from the same Normal distribution (see Sect. 2.3.1).

The marginal or posterior predictive distribution $f(y \mid x, H_1)$ is also a Normal distribution with mean equal to the posterior mean μ_x and variance equal to the sum of the posterior variance τ_x^2 and the population variance σ^2, that is,

$$(Y \mid x, H_1) \sim N(\mu_x, \tau_x^2 + \sigma^2). \tag{1.21}$$

The same arguments apply to the marginal or predictive distribution $f(y \mid H_2)$ in the denominator, which is a Normal distribution with mean equal to the prior mean μ and variance equal to the sum of the prior variance τ^2 and the population variance σ^2, that is,

(continued)

Example 1.2 (continued)

$$(Y \mid H_2) \sim N(\mu, \tau^2 + \sigma^2).$$ (1.22)

The Bayes factor can then obtained as follows:

$$
\begin{aligned}
BF &= \frac{N(y \mid \mu_x, \tau_x^2 + \sigma^2)}{N(y \mid \mu, \tau^2 + \sigma^2)} \\
&= \frac{(\tau_x^2 + \sigma^2)^{-1/2} \exp\left\{-\frac{1}{2} \frac{(y - \mu_x)^2}{\tau_x^2 + \sigma^2}\right\}}{(\tau^2 + \sigma^2)^{-1/2} \exp\left\{-\frac{1}{2} \frac{(y - \mu)^2}{\tau^2 + \sigma^2}\right\}}.
\end{aligned}
$$

Note that this can be easily extended to cases with multiple measurements $y = (y_1, \ldots, y_n)$ (see Sect. 3.3.1).

Note that the value of the measurements y and x may be expressed as a ratio of the marginal likelihoods in (1.17) and (1.18), that is,

$$
\begin{aligned}
BF &= \frac{\int f(y \mid \theta) f(x \mid \theta) \pi(\theta \mid H_1) d\theta}{f(x \mid H_1)} \times \frac{1}{f(y \mid H_2)} \\
&= \frac{\int f(y \mid \theta) f(x \mid \theta) \pi(\theta \mid H_1) d\theta}{\int f(x \mid \theta) \pi(\theta \mid H_2) d\theta \int f(y \mid \theta) \pi(\theta \mid H_2) d\theta},
\end{aligned}
$$ (1.23)

as $f(x \mid H_1) = f(x \mid H_2)$. If the recovered item and the control item come from the same source (i.e., hypothesis H_1 holds), then $\theta_y = \theta_x$, otherwise $\theta_y \neq \theta_x$ (i.e., hypothesis H_2 holds). If H_2 is true and hence the examined items come from different sources, the measurements can be considered independent. Note, however, that this is not necessarily the case. There are instances where the assumption of independence among measurements on control and recovered material under H_2 does not hold, and the BF will not simplify as in (1.23). See Linden et al. (2021) for a discussion about this issue in the context of questioned signatures.

The expression of the Bayes factor in (1.23) involves prior assessments about the unknown parameter θ, in terms of $\pi(\theta \mid H_i)$, as well as the likelihood function $f(\cdot \mid \theta)$. Thus, the Bayes factor cannot generally be regarded as a measure of the relative support to competing propositions provided by the data alone.

1.5.2   Score-Based Models

For some types of forensic evidence, the specification of a probability model for available data may be difficult. This is the case, for example, when the measurements are obtained using high-dimensional quantification techniques, e.g., for fingermarks or toolmarks (using complex sets of variables), in speaker recognition, or for traces such as glass, drugs or toxic substances that may be described by several chemical components. In such applications, a *feature-based* Bayes factor (Sect. 1.5.1) may not be feasible, and a *score-based* approach may represent a practicable (or even the only) available alternative. Broadly speaking, a score is a metric that summarizes the result of a forensic comparison of two items or traces, in terms of a single variable, representing a measure of similarity or difference (e.g., distance). Various distance measures can be used, such as *Euclidean* or *Manhattan* distance, see, e.g., Bolck et al. (2015).[14] One of the first proposals of score-based approaches in forensic science was presented in the context of forensic speaker recognition by Meuwly (2001).

Let $\Delta(\cdot)$ denote the function which assesses the degree of similarity between feature vectors x and y. The *similarity score* $\Delta(x, y)$ represents the evidence for which a Bayes factor is to be computed. The introduction of a score function for quantifying the similarities/dissimilarities between compared items allows one to reduce the dimensionality of the problem, while retaining the discriminative information as much as possible. For a score given by a distance, for example, one will expect a value close to zero if the features x and y relate to items from the same source. Vice versa, if the features x and y relate to items from different sources, one will expect a larger score, provided that there are differences between members in a population. The score-based Bayes factor (sBF) is

$$\text{sBF} = \frac{g(\Delta(x, y) \mid H_1, I)}{g(\Delta(x, y) \mid H_2, I)}, \tag{1.24}$$

where $g(\cdot)$ denotes the probability distribution associated with $\Delta(X, Y)$. For the sake of simplicity, the conditioning information I will be omitted hereafter.

For the Bayes factor in (1.24), one cannot reproduce the simplified expression that was derived in (1.16) for the feature-based Bayes factor. The score-based Bayes factor must be computed as the ratio of two probability density functions evaluated at the evidence score $\Delta(x, y)$, given the competing propositions H_1 and H_2. Since these two distributions are not generally available by default, the forensic examiner will generally try to derive a sBF using sampling distributions based on many scores produced under each of the two competing propositions. One way to compute the density of the score $\Delta(x, y)$ in the numerator is to generate many scores for comparisons between the known features x and the features y of other items known to come from the potential source assumed under H_1. The numerator can therefore be written as $\hat{g}(\Delta(x, y) \mid x, H_1)$, where $\hat{g}(\cdot)$ indicates that the distribution is constructed on the basis of relevant data (scores) produced for the case of interest.

---

[14] The score can also be interpreted as the inner product of two vectors (Neumann & Ausdemore, 2020).

In the denominator, it is assumed that the proposition H_2 is true, and x and y denote features of items that come from different sources. The challenge for the forensic examiner is that of selecting the most appropriate data for obtaining the distribution in the denominator. Note that there are different ways to address this question because, depending on the case at hand, it might be appropriate to condition on (i) the known source (i.e., pursuing a so-called *source-anchored* approach), (ii) the trace (i.e., *trace-anchored* approach), or (iii) none of these (i.e., *non-anchored* approach). This amounts to evaluating the score using the probability density distribution that is obtained by producing scores for comparisons between (i) the features x of the control item from the known source and features of items taken from randomly selected sources of the relevant population, (ii) the features y of the trace item and features of items taken from sources selected randomly in the relevant population, (iii) features of pairs of items taken from sources selected randomly in the relevant population (i.e., without using x and y). Formally, this amounts to defining the distribution in the denominator as follows:

(i) $\hat{g}(\Delta(x, y) \mid x, H_2)$,
(ii) $\hat{g}(\Delta(x, y) \mid y, H_2)$,
(iii) $\hat{g}(\Delta(x, y) \mid H_2)$.

See, e.g., Hepler et al. (2012) for a discussion of this topic.

Example 1.3 (Image Comparison) Consider a hypothetical case where the face of an individual is captured by surveillance cameras during the commission of a crime. Available screenshots are compared with the reference image(s) of a person of interest. For image comparison purposes, the evidence to be considered is a score given by the distance between the feature vectors x of the known reference and the evidential recording y (see Jacquet and Champod (2020) for a review). Consider the following competing propositions. H_1: The person of interest is the individual shown in the images of the surveillance camera, versus H_2: An unknown person is depicted in the image of the surveillance camera. To help specify the probability distribution of the score in the numerator, one can take several pairs of images from the person of interest to serve as pairs of questioned and reference items. To inform the probability distribution for the score in the denominator, conditioning on the reference item x (i.e., the images depicting the person of interest) can be justified as it may contain information that is relevant to the case and may be helpful for generating scores (Jacquet & Champod, 2020; Hepler et al., 2012). The distribution in the denominator can thus be computed using a *source-anchored* approach as in (i). The sBF can therefore be obtained as

(continued)

Example 1.3 (continued)

$$\text{sBF} = \frac{\hat{g}(\Delta(x, y) \mid x, H_1)}{\hat{g}(\Delta(x, y) \mid x, H_2)}.$$

In other types of forensic cases, conditioning on y in the denominator, case (ii), may be more appropriate. This represents an asymmetric approach to defining the distribution in the numerator and in the denominator.

Example 1.4 (Handwritten Documents) Consider a case involving handwriting on a questioned document. Handwriting features y on the questioned document are compared to the handwriting features x of a person of interest. The similarities and differences between x and y are measured by a suitable metric (score). To inform about the probability distribution of the scores in the numerator, one can take several draws of pairs of handwritten characters originating from the known source to serve as recovered and control items and to obtain scores from the selected draws. Under H_2, consideration of x is not relevant for the assessment. Note that here H_2 is the proposition according to which the person of interest is not the source of the handwriting on the questioned document, but someone else from the relevant population. It would then seem reasonable to construct the distribution for the denominator by comparing the features y of the questioned document with features x from items of handwriting of persons randomly selected from the relevant population of potential writers. This amounts to a *trace-anchored* approach as in situation (ii) defined above. In fact, for handwriting, the approach (i) would amount to discarding relevant information related to the questioned document. The sBF can therefore be obtained as

$$\text{sBF} = \frac{\hat{g}(\Delta(x, y) \mid x, H_1)}{\hat{g}(\Delta(x, y) \mid y, H_2)}.$$

In yet other cases, the distribution in the denominator may be obtained by comparing pairs of items drawn randomly from the relevant population, without conditioning on either x or y. In such cases, the alternative proposition H_2 is that the two compared items come from different sources.

Example 1.5 (Firearm Examination) Consider a case in which a bullet is found at a crime scene and a person carrying a gun is arrested. The extent of agreement between marks left by firearms on bullets can be summarized by a score or multiple scores. An example of a simple score is the concept of consecutive matching striations. To inform the distribution in the numerator, the scientist fires multiple bullets using the seized firearm. To inform the distribution in the denominator, the scientist fires and compares many bullets known to come from different guns (i.e., different relevant models). The distribution in the denominator can thus be computed using a *non-anchored* approach. The sBF can therefore be obtained as

$$\text{sBF} = \frac{\hat{g}(\Delta(x, y) \mid x, H_1)}{\hat{g}(\Delta(x, y) \mid H_2)}.$$

Note that this is a coarse approach in the sense that no consideration is given to general manufacturing features. Indeed, the amount and quality of striation on a bullet may depend on aspects such as the caliber and the composition (e.g., jacketed/non-jacketed bullets, etc.), hence a conditioning on y may be considered.

Another example for a non-anchored approach, in the context of fingermark comparison, can be found in Leegwater et al. (2017). An example will be presented in Sect. 3.3.4.

Note that the above considerations refer to so-called *specific-source* cases. In such cases, recovered material is compared to material from a known source. However, there are also other situations where the competing propositions are as follows:

H_1: The recovered and the control material originate from the *same* source.
H_2: The recovered and the control material originate from *different* sources.

For such *common-source* propositions, the sampling distributions under the competing propositions can be learned, under H_1, from many scores for known same-source pairs (with each pair drawn from a distinct source) and, under (H_2), from many scores for pairs known to come from different sources. The score-based BF in this case will account for the occurrence of the observed score under the competing propositions, but it does not account for the rarity of the characteristics of the trace.

While a score-based approach has the potential to reduce the dimensionality of the problem, the use of scores implies a loss of information because features y and x are replaced by a single score. Therefore, there is a trade-off to be found between the complexity of the original configuration of features and the performance of the score-metric, the choice of which requires a justification.

For a critical discussion of score-based evaluative metrics, see Neumann (2020) and Neumann and Ausdemore (2020). See also Bolck et al. (2015) for a discussion of feature- and score-based approaches for multivariate data.

1.6 Bayes Factor in the Investigative Setting

While the use of the Bayes factor for evaluative purposes is rather well established in both theory and practice, the focus on investigative settings still offers much room for original developments. In many forensic settings, especially in early stages of an investigation, it may be that no potential source is available for comparison. In such situations, it will not be possible to compare characteristics observed on recovered and reference materials, as would be the case in an evaluative setting (Sect. 1.5). Nevertheless, one can derive valuable information from the recovered material alone. Consider, for example, two populations denoted p_1 and p_2, respectively, and the following two propositions:

H_1:   The recovered item comes from population p_1 (e.g., a population of females).
H_2:   The recovered item comes from population p_2 (e.g., a population of males).

Denote by y the measurements on the recovered material known to belong to one of the two populations specified by the competing hypotheses, but it is not known which one. For such a situation, the Bayes factor measures the change produced by the measurements y on the recovered item in the odds in favor of H_1, as compared to H_2, when going from the prior to the posterior distribution.

Assume that a parametric statistical model $\{f(\cdot \mid \theta), \theta \in \Theta\}$ is suitable for the data at hand. The problem of discriminating between two populations can then be treated as a problem of comparing statistical hypotheses, assuming that the probability distribution for the measurements on the recovered material (under each hypothesis) is of a given form. Consider, first, the situation where the parameters characterizing the two populations are known, that is, $\theta = \theta_1$ if the recovered item comes from population p_1 and $\theta = \theta_2$ if the recovered item comes from population p_2. Formally, this amounts to specifying the probability distributions $f(y \mid \theta_1)$ and $f(y \mid \theta_2)$, respectively. The posterior probability of the competing propositions can be computed as in (1.5) and the Bayes factor simplifies to a ratio of likelihoods as in (1.6).

Example 1.6 (Fingermark Examination) Consider a case involving a single fingermark of unknown source. The fingerprint examiner seeks to help with the question of whether the mark comes from a man or woman. Thus, for investigative purposes, the following two propositions are of interest:

(continued)

Example 1.6 (continued)

H_1:   The fingermark comes from a man.

H_2:   The fingermark comes from a woman.

A type of data that can be acquired from fingermarks is ridge width, summarized in terms of the ridge count per surface in mm^2. See, for example, Appendix A of Champod et al. (2016) for a summary of different data collections. Consider ridge density, which was found to vary as a function of sex (i.e., women tend have narrower ridges than men), but also between populations. Suppose that normality represents a reasonable assumption for ridge density so that the probability distribution for available measurements can be considered Normal $N(\theta_i, \sigma_i^2)$, with the unknown mean θ being equal to θ_i and the variance σ^2 being equal to σ_i^2 if H_i is true. Given H_1, the measurements y thus have a probability distribution $N(\theta_1, \sigma_1^2)$ and given H_2 a probability distribution $N(\theta_2, \sigma_2^2)$.

The posterior probability of the competing propositions can be computed as in (1.5), and the Bayes factor simplifies to a likelihood ratio as in (1.6), that is,

$$BF = \frac{N(y \mid \theta_1, \sigma_1^2)}{N(y \mid \theta_2, \sigma_2^2)}.$$

Generally, however, the parameters, or some of the parameters, characterizing the two distributions are unknown and a pair of probability density distributions will be introduced to model this uncertainty. As a consequence, the Bayes factor will also depend on prior assumptions and will not simplify to a likelihood ratio. Consider the case where parameters θ_i are continuous and take values in the parameter space Θ_i. A prior distribution $\pi(\theta_i \mid p_i)$ must be specified, with $\theta_i \in \Theta_i$ and p_i representing the population of interest. A marginal distribution for each population p_i can be computed as in (1.2),

$$f_{H_i}(y) = \int_{\Theta_i} f(y \mid \theta_i)\pi(\theta_i \mid p_i)d\theta_i \qquad (1.25)$$

and the Bayes factor will take the form of a ratio of marginal likelihoods as in (1.3), that is,

$$BF = \frac{f_{H_1}(y)}{f_{H_2}(y)}. \qquad (1.26)$$

Example 1.7 (Fingermark Examination—Continued) Recall Example 1.6 where a Normal probability distribution was assumed for the measured ridge density on a fingermark, with variance known and equal to σ_i^2. A conjugate prior distribution may be introduced for the population mean θ_i, say $\theta_i \sim N(\mu_i, \tau_i^2)$. The marginal likelihoods are still Normal with mean equal to the prior mean μ_i and variance equal to the sum of the prior variance τ_i^2 and the population variance σ_i^2. The Bayes factor therefore is

$$\text{BF} = \frac{N(y \mid \mu_1, \tau_1^2 + \sigma_1^2)}{N(y \mid \mu_2, \tau_2^2 + \sigma_2^2)}.$$

The same idea can be extended to the case where both the mean and the variance are unknown. This will be addressed in Sect. 4.3.2.

The Bayes factor thus depends on the prior *assumptions* about parameters characterizing each population. This must not be confused, as noted earlier, with prior probabilities for competing propositions. The latter will form the prior *odds* which will be multiplied by the Bayes factor to compute the posterior odds

$$\frac{\Pr(H_1 \mid y)}{\Pr(H_2 \mid y)} = \frac{f_{H_1}(y)}{f_{H_2}(y)} \times \frac{\Pr(H_1)}{\Pr(H_2)}.$$

The Bayesian approach for discriminating between two propositions regarding population membership can be easily generalized to the case where there are any number k (>2) of competing mutually exclusive propositions. Let $H_1, \dots, H_k$ denote k propositions and denote by y the observation to be evaluated. The propositions of interest can be defined as follows:

H_1:   The recovered item comes from population 1 (p_1).
H_2:   The recovered item comes from population 2 (p_2).
  $\vdots$
H_k: The recovered item comes from population k (p_k).

Example 1.8 (Questioned Documents) Consider a case involving questioned documents where the issue of interest is which of three printing machines has been used to print a questioned document. Propositions of interest are:

H_1:   The questioned documents have been printed with printer 1.
H_2:   The questioned documents have been printed with printer 2.

(continued)

Example 1.8 (continued)

H_3: The questioned documents have been printed with printer 3.

After having specified a Bayesian statistical model for each proposition (i.e., a probability distribution for the available measurements and a prior distribution for the unknown parameters), the marginal likelihoods $f_{H_i}(y)$, $i = 1, 2, 3$, characterizing propositions H_1, H_2, and H_3, can be obtained as in (1.25).

Occasionally, cases involve multiple propositions. Imagine a case involving DNA findings, such as bloodstains recovered on a crime scene, with the reported profile being compared against the profile of a person of interest. The defense argues that the bloodstain does not come from the person but from either a relative (e.g., a brother) or an unknown person. A question that may arise in such a case is how to evaluate and report results, because the Bayes factor involves pairwise comparisons. One option is to report only the marginal likelihoods $f_{H_i}(y)$, even if they may not be easy to interpret. Alternatively, one may report a scaled version $f^*_{H_i}(y)$ as suggested by Berger and Pericchi (2015), that is,

$$f^*_{H_i}(y) = \frac{f_{H_i}(y)}{\sum_{j=1}^{k} f_{H_j}(y)}. \tag{1.27}$$

This expression will be much easier to interpret, because the scaled likelihoods $f^*_{H_i}(y)$ sum up to 1. Generally, prior probabilities $\Pr(H_i)$ may vary between recipients of such reports, but the posterior probability can be easily computed as

$$\Pr(H_i \mid y) = \frac{\Pr(H_i) f^*_{H_i}(y)}{\sum_{j=1}^{k} \Pr(H_j) f^*_{H_j}(y)}, \qquad i = 1, \ldots, k$$

followed, if required, by classification of the recovered material in the population with the highest posterior probability. Note that reporting the scaled version in (1.27) is equivalent to assuming equal prior probabilities for competing propositions. In fact, if $\Pr(H_i) = \frac{1}{k}$, $i = 1, \ldots, k$, then it can easily be shown that

$$\Pr(H_i \mid y) = \frac{f^*_{H_i}(y)}{\sum_{j=1}^{k} f^*_{H_j}(y)} = f^*_{H_i}(y), \qquad i = 1, \ldots, k,$$

as $\sum_{j=1}^{k} f^*_{H_j}(y) = 1$.

The analyst may also consider the possibility of summarizing several propositions into one, in order to produce a comparison between two propositions regarding population membership. One of these propositions will be composite. Let $\bar{H}_1 =$

$H_2 \cup \cdots \cup H_k$. Starting from k possible populations from which the recovered material may come from, a pair of competing propositions of interest may thus be formulated as follows:

H_1:   The recovered item comes from population 1 (p_1).
$\bar{H}_1$:   The recovered item comes from one of the other populations ($p_2, \ldots, p_k$).

The marginal likelihood $f_{H_1}(y)$ characterizing proposition H_1 is obtained as in (1.25), while the marginal likelihood under $\bar{H}_1$ is

$$f_{\bar{H}_1}(y) = \sum_{i=2}^{k} \Pr(p_i) \int_{\Theta_i} f(y \mid \theta_i) \pi(\theta_i \mid p_i) d\theta_i,$$

with $\sum_{i=1}^{k} \Pr(p_i) = 1$. The Bayes factor expressing the value of y for comparing H_1 against $\bar{H}_i$ becomes

$$\mathrm{BF} = \frac{f_{H_1}(y) \sum_{i=2}^{k} \Pr(p_i)}{f_{\bar{H}_1}(y)}. \tag{1.28}$$

The posterior odds become

$$\frac{\Pr(H_1 \mid y)}{\Pr(\bar{H}_1 \mid y)} = \frac{f_{H_1}(y) \Pr(p_1)}{f_{\bar{H}_1}(y)},$$

(Aitken et al., 2021, p. 643).

Example 1.9 (Questioned Documents—Continued) Consider the following propositions:

H_1:   The questioned documents have been printed with printer 1.
$\bar{H}_1$:   The questioned documents have been printed with printer 2 or with printer 3.

The marginal likelihood characterizing proposition H_1 is

$$f_{H_1}(y) = \int_{\Theta_1} f(y \mid \theta_1) \pi(\theta_1 \mid p_1) d\theta_1.$$

The marginal likelihood characterizing proposition $\bar{H}_1$ will become

$$f_{\bar{H}_1}(y) = \Pr(p_2) \int_{\Theta_2} f(y \mid \theta_2) \pi(\theta_2 \mid p_2) d\theta_2$$

(continued)

Table 1.1 Scale for verbally expressing support provided by the observations for one hypothesis over an alternative adapted from Jeffreys (1961)

BF	Evidence in favor of H_1
1 to 3.2	Not worth more than a bare mention
3.2 to 10	Substantial
10 to 100	Strong
>100	Decisive

Table 1.2 Verbal scale for expressing evidential value, in terms of the Bayes factor, in support of the prosecution's proposition over the alternative (defense) proposition (Willis et al., 2015)

Value of the BF	Verbal equivalent: The forensic findings …
1	Do not support one hypothesis over the other
2 to 10	Provide weak support (for the first hypothesis relative to the alternative)
10 to 100	Provide moderate support (*idem*)
100 to 1000	Provide moderately strong support (*idem*)
1000 to 10,000	Provide strong support (*idem*)
10,000 to 1,000,000	Provide very strong support (*idem*)
1,000,000 and above	Provide extremely strong support (*idem*)

Example 1.9 (continued)

$$+ \Pr(p_3) \int_{\Theta_3} f(y \mid \theta_3)\pi(\theta_3 \mid p_3)d\theta_3,$$

and the Bayes factor can be obtained as in (1.28).

1.7   Bayes Factor Interpretation

The Bayes factor is a coherent measure of the change in support that the findings provide for one hypothesis against a given alternative (Jeffrey, 1975). Table 1.1 shows a guide for expressing Bayes factors verbally, following Jeffreys (1961). A historical review is presented in Aitken and Taroni (2021).

The verbal equivalent must express a degree of support for one of the propositions relative to an alternative and is defined from ranges of Bayes factor values. Qualitative interpretations of the Bayes factor have also been proposed in the context of forensic science (Evett, 1987, 1990; Evett et al., 2000; Nordgaard et al., 2012; Willis et al., 2015). Table 1.2 summarizes an example of a scale given in the ENFSI Guideline for Evaluative Reporting in Forensic Science (Willis et al., 2015), inspired by the scale proposed by Nordgaard et al. (2012). Users of these scales must be aware that labelling several Bayes factor apportionments offers a broad descriptive statement about standards of evidence in scientific investigation and not a calibration of the Bayes factor (Kass, 1993). See, e.g., Ramos and Gonzalez-Rodriguez (2013), van Leeuwen and Brümmer (2013) and Aitken et al. (2021) for an account of calibration as a measure of performance of BF computation methods.

Moreover, it is important to note that the choice of a reported verbal equivalent is based on the magnitude of the Bayes factor and not the reverse. Marquis et al. (2016) present a discussion on how to implement a verbal scale in a forensic laboratory, considering benefits, pitfalls, and suggestions to avoid misunderstandings.

It is worth to reiterate that a Bayes factor represents *a measure of change in support* rather than *a measure of support*, though the two expressions may be perceived as equivalent. In fact, the Bayes factor can be shown to be a non-coherent measure of support: a small Bayes factor means that the data will lower the probability of the hypothesis of interest relative to its value *prior* to considering the evidence, but it does not imply that the probability of this hypothesis is low. The Bayes factor measures the *change* produced in the odds, thus providing a measure of whether the available findings have increased or decreased the odds in favor of one proposition compared to the alternative (Bernardo & Smith, 2000).

1.8 Computational Aspects

The computation of Bayes factors can be challenging, especially when the marginal likelihoods in the numerator and in the denominator (1.2) involve integrals that do not have an analytical solution. Several methods have been proposed to address this complication. See Kass and Raftery (1995) and Han and Carlin (2001) for a review.

Consider the following general expression for the marginal likelihood:

$$f(x) = \int f(x \mid \theta)\pi(\theta)d\theta. \tag{1.29}$$

If the likelihood $f(x \mid \theta)$ and the prior $\pi(\theta)$ are not family conjugates, then an analytical solution may not be available. But suppose that it is possible to draw values from the prior distribution $\pi(\cdot)$. The integral in (1.29) can then be approximated by Monte Carlo methods as

$$\hat{f}_1(x) = \sum_{i=1}^{N} f(x \mid \theta^{(i)})/N, \tag{1.30}$$

where $\theta^{(i)}$, $i = 1, \ldots, N$, are N independent draws from $\pi(\cdot)$. This is the average of the likelihood of the sampled values. An example will be provided in Sect. 2.2.2 (Example 2.3).

This simulation process can be rather inefficient when the posterior distribution is concentrated, relative to the prior, as most of the $\theta^{(i)}$ will have a small likelihood and the estimate $\hat{f}_1(x)$ in (1.30) may be dominated by a few values with large likelihood. The precision of the Monte Carlo integration can be improved by importance sampling (Kass & Raftery, 1995). Moreover, statistical packages (e.g., in R) allow one to sample from a certain number of distributions.

Importance sampling as well as other Monte Carlo tools may help to overcome such difficulties as there is no need for the distribution $\pi(\theta)$ to be available in closed form. Consider any manageable density $\pi^*(\theta)$ from which it is feasible to sample. The integral in (1.29) can then be approximated by *importance sampling* as

$$\hat{f}_2(x) = \frac{\sum_{i=1}^{N} w_i f(x \mid \theta^{(i)})}{\sum_{i=1}^{N} w_i}, \tag{1.31}$$

where $\theta^{(i)}$ are independent draws from $\pi^*(\theta)$ and are weighted by importance weights $w_i = \pi(\theta^{(i)})/\pi^*(\theta^{(i)})$. The function $\pi^*(\theta)$ is known as *importance sampling function* (e.g., Geweke, 1989). An example will be provided in Sect. 2.2.2 (Example 2.3).

In the case where $\pi^*(\theta)$ is taken to be the posterior density $\pi(\theta \mid x) = \pi(\theta)f(x \mid \theta)/f(x)$, the use of this expression in (1.31) yields the harmonic mean of the sampled likelihood values as an estimate for the marginal likelihood $f(x)$:

$$\hat{f}_3(x) = \left[\frac{1}{N} \sum_{i=1}^{N} \frac{1}{f(x \mid \theta_i)} \right]^{-1}.$$

Note that, whatever method is used, the output of such a simulation procedure is an approximation that must be handled carefully. Notwithstanding, it is worth pointing out that while the *Monte Carlo estimate* is not exact, the Monte Carlo error (e.g., $f(x) - \hat{f}_1(x)$) can be very small if a sufficiently large number of draws are generated. A study of Monte Carlo errors for the quantification of the value of forensic evidence is provided by Ommen et al. (2017).

Many practical problems require more advanced techniques based on Markov chain Monte Carlo methods (MCMC) to overcome computational hurdles. The general idea behind these methods is to sample recursively values $\theta^{(i)}$ from some transition distribution that depends on the previous draw $\theta^{(i-1)}$ in such a way that at each step of the iteration process, we expect to draw from a distribution that becomes closer (i.e., converges) to the target posterior distribution $\pi(\theta \mid x)$. This means that, for many iterations, $\theta^{(i)}$ is approximately distributed according to $\pi(\theta \mid x)$ and can be used like the output of a Monte Carlo simulation algorithm. To avoid the effect of starting values, the first set of iterations is generally discarded (this is called the *burn in* period), and the simulated values beyond the first n_b iterations

$$\theta^{(n_b+1)}, \ldots, \theta^{(N)}$$

are taken as draws from the target posterior distribution. The Gibbs sampling algorithm is a well-known method to construct a chain with these features. Suppose that the parameter vector can be decomposed into several components, say $\theta = (\theta_1, \ldots, \theta_p)$, and let $\pi(\theta_j \mid \theta_{-j}^{(i-1)})$ denote the so-called full conditional distribution, that is the conditional distribution of θ_j at step (i) given all the other components, say θ_{-j}, at the previous step $(i-1)$

$$\theta_{-j}^{(i-1)} = (\theta_1^{(i-1)}, \ldots, \theta_{j-1}^{(i-1)}, \theta_{j+1}^{(i-1)}, \ldots, \theta_p^{(i-1)}).$$

For many problems, it is possible to sample easily from the conditional distributions, as is the case when distributions are conjugate. The Gibbs sampling algorithm works as follows: start with an arbitrary value $\theta^{(0)} = (\theta_1^{(0)}, \ldots, \theta_p^{(0)})$ and generate $\theta_j^{(i)}$ at each iteration according to the conditional distribution given the current values $\theta_{-j}^{(i-1)}$. Examples will be given in Sects. 3.4.1.3 (Example 3.14) and 3.4.3 (Example 3.16.)

Whenever it is not possible to decompose the joint distribution in manageable conditionals, one can implement an alternative approach, the Metropolis–Hastings (M–H) algorithm (e.g. Gelman et al., 2014). This algorithm can be summarized as follows. Start with an arbitrary value $\theta^{(0)} = (\theta_1^{(0)}, \ldots, \theta_p^{(0)})$ and generate $\theta_j^{(i)}$ at each iteration, as follows:

1. Draw a proposal value θ_j^{prop} form a density $q(\theta_j^{(i-1)}, \theta_j^{\mathrm{prop}})$, called *candidate generating density*.
2. Compute a probability of acceptance as follows:

$$\alpha\left(\theta_j^{(i-1)}, \theta_j^{\mathrm{prop}}\right) = \min\left\{ \frac{\pi\left(\theta_j^{\mathrm{prop}}\right) q\left(\theta_j^{\mathrm{prop}}, \theta_j^{(i-1)}\right)}{\pi\left(\theta_j^{(i-1)}\right) q\left(\theta_j^{(i-1)}, \theta_j^{\mathrm{prop}}\right)} \right\}. \qquad (1.32)$$

3. Accept the proposed value θ_j^{prop} with probability $\alpha\left(\theta_j^{(i-1)}, \theta_j^{\mathrm{prop}}\right)$, and set $\theta_j^{(i)} = \theta_j^{\mathrm{prop}}$; otherwise, reject the proposed value and set $\theta_j^{(i)} = \theta_j^{(i-1)}$.

Note that if the candidate generating function is symmetric (e.g., a Normal probability density), the acceptance probability in (1.32) simplifies to

$$\alpha\left(\theta_j^{(i-1)}, \theta_j^{\mathrm{prop}}\right) = \min\left\{ \frac{\pi\left(\theta_j^{\mathrm{prop}}\right)}{\pi\left(\theta_j^{(i-1)}\right)} \right\}.$$

The performance of an MCMC algorithm can be monitored by inspecting graphs and computing diagnostic statistics. Such exploratory analysis is fundamental for assessing convergence to the posterior distribution. An example will be given in Sect. 2.2.2 (Example 2.6).

The output of the MCMC algorithm can be used to provide the marginal likelihood that is needed for the numerator and the denominator of the Bayes factor, as proposed by Chib (1995) for a Gibbs sampling algorithm and by Chib and Jeliazkov (2001) for an M–H algorithm. The key idea is to obtain the marginal likelihood $f(x)$ by a direct application of Bayes theorem since it can be seen as the normalizing constant of the posterior density

$$f(x) = \frac{f(x \mid \theta^*)\pi(\theta^*)}{\pi(\theta^* \mid x)}, \tag{1.33}$$

where θ^* is a parameter value with high posterior density. Note that (1.33) is valid for any parameter value $\theta \in \Theta$. The likelihood $f(x \mid \theta)$ and the prior density $\pi(\theta)$ can be directly computed at a given parameter point θ^*. The posterior density $\pi(\theta \mid x)$ is unavailable in closed form, but it can be approximated by using the output of the Gibbs sampling. Consequently, the marginal likelihood can be approximated as

$$\hat{f}(x) = \frac{f(x \mid \theta^*)\pi(\theta^*)}{\hat{\pi}(\theta^* \mid x)}. \tag{1.34}$$

Examples will be given in Sects. 3.4.1.3 (Example 3.14) and 3.4.3 (Example 3.16).

This short overview of computational tools is not intended to be exhaustive. There are instances, for example when dealing with high-dimensional distributions, where the simulation process is very slow, giving rise to inefficiencies in the behavior of the Gibbs sampler or Metropolis algorithm. An alternative solution is given by the Hamiltonian Monte Carlo (HMC) method, where the proposal distribution is not centered on the current position of the chain and changes depend on the current position of the chain. This allows one to obtain more promising candidate values, avoiding to get stuck in a very slow exploration of the target distribution and therefore to move much more rapidly (Neal, 1996). As in any Metropolis algorithm, the HMC proceeds by a series of iterations, though it requires more efforts in terms of programming and tuning. The user can refer to a computer program called Stan (sampling through adaptive neighborhoods) to directly apply the Hamiltonian Monte Carlo method. The reader can refer to Gelman et al. (2014) and Stan Development Team (2021) for instructions and examples. A complete picture of basic and more advanced methods of Bayesian computation can be found, e.g., in Gelman et al. (2014), Marin and Robert (2014), and Robert and Casella (2010). The reader can also refer to Han and Carlin (2001) and to Friel and Pettitt (2008) for a review of methods to compute BFs.

In all examples in this book, dealing with the Gibbs sampler and the Metropolis–Hastings algorithm, we will directly program the computations in R. Other open-source programs however exist that can be used to build Markov chain Monte Carlo sampler, such as Stan or Jags (Just another Gibbs sampler, https://mcmc-jags.sourceforge.io/). They both can interact with R (see libraries RStan, rjags and runjags). Further examples can be found in Albert (2009) and Kruschke (2015).

1.9   Bayes Factor and Decision Analysis

The Bayes factor provides a coherent and quantitative way for relating probabilities for states of nature, before information is obtained, to probabilities given information that has become available. A subsequent step, the choice between different hypotheses, represents a problem of decision-making (Lindley, 1985). For the purpose of illustration, consider the simple and regularly encountered case where only two hypotheses are of interest, say H_1 and H_2. The two hypotheses represent the list of, more generally, n exclusive and exhaustive uncertain events (also called *states of nature*) and denote the entirety of nature. The decision space is the set of all possible actions, here decisions d_1 and d_2, where decision d_i can be formalized as the acceptance of hypothesis H_i. The decision problem can be expressed in terms of a decision matrix (see Table 1.3) with C_{ij} denoting the consequence of deciding d_i when hypothesis H_j is true. Decision d_i is called "correct" if hypothesis H_j is true and $i = j$. Conversely decision d_i is not correct if hypothesis H_j is true and $i \neq j$, i.e., $H_{\neg i}$ is true. When choosing between competing hypotheses, one takes preferences among decision consequences into account, in particular among adverse outcomes. This aspect is formalized by introducing a measure for expressing the decision maker's relative desirability, or undesirability, of the various decision consequences. To measure the undesirability of consequences on a numerical scale, one can introduce a loss function $L(\cdot)$, where $L(C_{ij})$ denotes the loss that one assigns to the outcome of deciding d_i when hypothesis H_j is true.

If it can be agreed that a correct decision represents neither a loss nor a gain, the loss function for a two-action problem can be described with a two-way table that contains zeros for the losses $L(C_{ij})$, $i = j$, and the value l_i for $L(C_{ij})$, $i \neq j$. Such a "$0 - l_i$" loss function is shown in Table 1.4, where $l_i = L(d_i, H_{\neg i})$ denotes the loss one incurs whenever decision d_i is a wrong decision.

The relative (un-)desirability of available decisions can be expressed by their *expected loss* $EL(\cdot)$, computed as

Table 1.3 Decision matrix with d_1 and d_2 denoting the possible actions, H_1 and H_2 denoting the states of nature, and C_{ij} denoting the consequence of deciding d_i when hypothesis H_j is true

	H_1	H_2
d_1	C_{11}	C_{12}
d_2	C_{21}	C_{22}

Table 1.4 The "$0 - l_i$" loss function for a decision problem with d_1 and d_2 denoting the possible actions, H_1 and H_2 denoting the states of nature, and l_i denoting the loss associated with adverse decision consequences

	H_1	H_2
d_1	0	l_1
d_2	l_2	0

$$\mathrm{EL}(d_i \mid x) = \underbrace{\mathrm{L}(d_i, H_i)}_{0}\underbrace{\Pr(H_i \mid x)}_{\alpha_i} + \underbrace{\mathrm{L}(d_i, H_{\neg i})}_{l_i}\underbrace{\Pr(H_{\neg i} \mid x)}_{\alpha_{\neg i}}$$

$$= l_i \alpha_{\neg i},$$

where x denotes the observation or a series of measurements and $\alpha_{\neg i}$ denotes the (posterior) probability of the event $H_{\neg i}$ given x. The formal Bayesian decision criterion is to accept hypothesis H_1 if the expected loss of the decision to accept H_1 is smaller than the expected loss of rejecting it, that is, if the (posterior) expected loss of decision d_1 is smaller than the (posterior) expected loss of decision d_2:

$$\mathrm{EL}(d_1 \mid x) < \mathrm{EL}(d_2 \mid x)$$

$$l_1 \alpha_2 < l_2 \alpha_1. \tag{1.35}$$

When rearranging the terms in (1.35) to $\alpha_1/\alpha_2 > l_1/l_2$, and dividing both sides by the prior odds π_1/π_2, the Bayes decision criterion states that accepting H_1 is the optimal decision whenever

$$\frac{\alpha_1/\alpha_2}{\pi_1/\pi_2} > \frac{l_1/l_2}{\pi_1/\pi_2} = c.$$

This is equivalent to accepting H_1 whenever the Bayes factor in favor of this proposition is larger than a constant c determined by the prior odds and the loss ratio. Given a set of observations, the Bayes factor is computed and, depending on whether or not it exceeds a given threshold, the decision maker chooses between the members in the list of states of nature (here H_1 and H_2). Examples will be given in Chap. 3 in the context of inference of source (Sect. 3.3.3) and in Chap. 4 in the context of classification (Sects. 4.2.2 and 4.4.1.2). An extended review of elements of decision analysis in forensic science can be found in Taroni et al. (2021b).

This decision criterion is simple and intuitive, yet it poses challenges. For example, the requirement to choose a prior probability for the two hypotheses may be discomforting, because there is no ad hoc recipe for this purpose. In principle, probabilities are personal, since they depend on one's knowledge (Lindley, 2014). They may change as information changes and may vary among individuals. For example, a given hypothesis may be considered almost true by one individual, but far less probable by someone else. The fact that different individuals with different knowledge bases may specify different probabilities for the same event, provided that they are accompanied with a justification, is not a problem in principle (Lindley, 2000). The only strict requirement to which probability assignments ought to conform is coherence (de Finetti, 2017). Coherence has the normative role of encouraging people to make careful assignments based on their personal knowledge. This can be operationally supported by the concept of scoring rules. See, for example, Biedermann et al. (2013, 2017a) for a discussion of scoring rules in the context of forensic science.

The same viewpoint applies to utility and loss functions, which may be difficult to specify. A "correct" utility (or loss) function does not exist, because preference structures are personal. Adverse decision consequences may be considered more or less undesirable, depending on the background, the context and the decision maker's objectives (e.g., Taroni et al., 2010). Moreover, the loss function does not need to have constant values, such as the "$0 - l_i$" loss function introduced above. More general loss functions treat the loss as a function of the severity of the consequences. Examples will be given in Chap. 2 regarding inference and decision about a proportion (Sect. 2.2.3) and about a mean (Sect. 2.3.3).

Note that, in the context here, the terms "personal" and "subjective" do not mean that the theory is arbitrary, unjustified or groundless (Biedermann et al., 2017b; Taroni et al., 2018). There are various devices for the sound elicitation of probabilities and the measurement of the value of decision consequences (Lindley, 1985). What matters in a situation in which a decision maker is asked to make a choice among alternative courses of action that have uncertain consequences is that the behavior is one that can be qualified as rational. This includes, in particular, a coherent specification of the loss function, reflecting personal preferences among consequences in terms of desirability or undesirability.

This formal decision-analytic approach provides decision criteria that (i) are based on clearly defined concepts, (ii) promote rational decision-making under uncertainty, and (iii) make a clear distinction between the evaluation of the strength of evidence (as given by the Bayes factor), which is the domain of the forensic scientist, and the specification of the threshold with which the Bayes factor is compared, i.e., the ratio between the loss ratio and the prior odds. The latter lies in the domain of the recipient of expert information, such as investigative authorities and members of the judiciary.

1.10   Choice of the Prior Distribution

Bayesian model builders may encounter various difficulties. One of them is the choice of the prior distribution. Bayes theorem does not specify how one ought to define the prior distribution. The chosen prior distribution should, however, suitably reflect one's prior beliefs. In this context, so-called vague or non-informative prior distributions may help to find a broad consensus. However, it is important to keep in mind that even a "non-informative" prior distribution effectively conveys a well-defined opinion, i.e., that probabilities spread uniformly over the parameter space (de Finetti, 1993a). In contrast to this, personal or so-called informative priors aim at encoding available prior knowledge. Whenever feasible, it is advantageous to choose a member of the class of conjugate distributions, that is, a family of prior distributions such that for any prior in this family and a particular probability distribution, the corresponding posterior distribution will be in the same family. For example, the beta distribution and the binomial distribution are said to be conjugate in this sense. Several examples will be provided throughout this book.

Table 1.5   Some common conjugate prior distribution families

Probability distribution	Conjugate prior distribution
Binomial:	Beta:
$f(x \mid \theta) = \text{Bin}(n, \theta)$	$\pi(\theta) = \text{Be}(\alpha, \beta)$
Multinomial:	Dirichlet:
$f(x_1, \ldots, x_k \mid \theta_1, \ldots, \theta_k) = \text{Mult}(n, \theta_1, \ldots, \theta_k)$	$f(\theta_1, \ldots, \theta_k) = \text{Dir}(\alpha_1, \ldots, \alpha_k)$
Poisson:	Gamma:
$f(x \mid \lambda) = \text{Pn}(\lambda)$	$\pi(\lambda) = \text{Ga}(\alpha, \beta)$
Normal (known variance):	Normal:
$f(x \mid \theta, \sigma^2) = \text{N}(\theta, \sigma^2)$	$\pi(\theta) = \text{N}(\mu, \tau^2)$
Normal (known mean):	Inverse Gamma:
$f(x \mid \theta, \sigma^2) = \text{N}(\theta, \sigma^2)$	$\pi(\sigma^2) = \text{IG}(\alpha, \beta)$

Table 1.5 provides a list of some common families of conjugate distributions. A more extensive list can be found in Bernardo and Smith (2000). Despite such smooth technical options, eliciting a prior distribution may not be easy.

First, it may be that none of the standard parametric families mentioned above is suitable to describe one's prior degree of belief. There may be circumstances where multimodal priors may better reflect the available knowledge, and mixture priors would be more convenient (see e.g. Taroni et al., 2010). Another option is to specify prior beliefs over a selection of points and then interpolate between them (Bolstad & Curran, 2017). More generally, there may be cases where the choice of a conjugate prior is not appropriate as it does not properly reflect available knowledge. If this is the case, the application of Bayes theorem may lead to a posterior distribution that is analytically intractable. Such situations require the implementation of computational tools as described in Sect. 1.8.

Second, practitioners will immediately realize that even if the choice of a given standard parametric family may appear justifiable, they will still need to choose a member from the selected family. Stated otherwise, they will need to fix the hyperparameters of the prior distribution in a way that the resulting shape will reasonably reflect their knowledge. Assume that practitioners are in a situation where, based on their experience in the field, they can summarize and translate their prior beliefs into a numerical value for the prior mean, say m, and into a numerical value for the prior standard deviation, say s. They can then find the values of the parameters that specify a prior distribution that reflects the assessed prior location and prior dispersion, respectively. For example, suppose that the parameter of interest, θ, is a proportion and that a beta prior distribution is chosen to model prior uncertainty, i.e., $\theta \sim \text{Be}(\alpha, \beta)$. The problem then is how to choose α and β. If one can specify a value m for the prior mean and a value s for the prior standard deviation, that is the two values describing the location and the shape of the prior distribution, one can elicit the hyperparameters α and β by relating the assessed prior mean and prior variance to the prior moments of a beta distributed random variable, that is,

$$m = \frac{\alpha}{\alpha + \beta} \tag{1.36}$$

$$s^2 = \frac{\alpha\beta}{(\alpha + \beta + 1)(\alpha + \beta)^2}. \tag{1.37}$$

The hyperparameters of the beta prior can then be obtained by solving the two equations in (1.36) and (1.37) for α and β

$$\alpha = m \left[\frac{m(1-m)}{s^2} - 1 \right] \tag{1.38}$$

$$\beta = (1-m) \left[\frac{m(1-m)}{s^2} - 1 \right]. \tag{1.39}$$

It is advisable to inspect the prior distribution thus elicited. Producing a graphical representation can help examine whether the shape of the distribution reasonably reflects one's prior beliefs. Moreover, the so-called *equivalent sample size n_e* should be calculated in order to examine the reasonableness of the amount of information that underlies the proposed prior; one should make sure that it is not unrealistically high (Bolstad & Curran, 2017). Stated otherwise, one should examine whether the information that is conveyed by the prior is equivalent, at least roughly, to the information that would be obtained by collecting a sample of equivalent size n_e. For example, consider a random sample $(X_1, \ldots, X_{n_e})$ of size n_e, providing the same information that is conveyed by the prior. The sample mean $\bar{X} = \frac{1}{n_e} \sum_{i=1}^{n_e} X_i$ should have, at least roughly, the same location and the same dispersion as the prior.

For the beta-binomial case, the equivalent sample size n_e can be obtained by relating the moments of the beta prior to the corresponding moments characterizing a random sample of size n_e from a Bernoullian population with probability of success θ:

$$\frac{\alpha}{\alpha + \beta} = \theta \tag{1.40}$$

$$\frac{\alpha\beta}{(\alpha + \beta + 1)(\alpha + \beta)^2} = \frac{\theta(1-\theta)}{n_e}. \tag{1.41}$$

Solving for n_e, one obtains $n_e = \alpha + \beta + 1$. If this is felt to be unrealistic, then one should revise one's prior assessments, increase the dispersion and recalculate the prior. Otherwise, one might specify too much information about the proportion θ relative to the amount of information provided by a sample of size n_e.

Example 1.10 (Elicitation of a Beta Prior) Suppose that a prior distribution needs to be elicited for the proportion θ of non-counterfeit merchandise (e.g., medicines) in a target population. It is thought that the distribution is centered around 0.8 with a standard deviation equal to 0.1. Parameters α and β can be elicited as in (1.38) and (1.39)

```
> m=0.8
> s=0.1
> a=m*(m*(1-m)/s^2-1)
> b=(1-m)*(m*(1-m)/s^2-1)
> c(a,b)

[1] 12   3
```

Figure 1.1 shows the elicited beta prior Be(12, 3).

```
> plot(function(x) dbeta(x,a,b),0,1,xlab=expression
+ (paste(theta)),ylab=expression(paste(pi)*
+ paste('(')*paste(theta)*paste(')')))
```

The equivalent sample size is 12+3+1=16. This is the size of the sample that should be available in terms of information that is equivalent to that conveyed by the elicited prior.

Fig. 1.1 Prior distribution Be(12, 3) over θ in Example 1.10

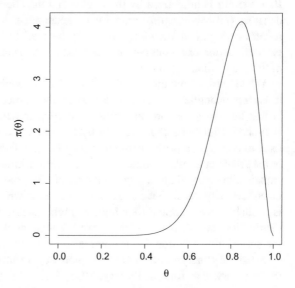

An objection to this procedure might be that while specifying a value for the location of the prior may be feasible, this may not necessarily be so for the dispersion. In many cases, the available prior knowledge takes the form of a realization $(x_1, \ldots, x_n)$ of a random sample of size n from a previous experiment. In this case, it is sufficient to solve (1.40) and (1.41) with respect to α and β for this sample size n:

$$\alpha = p(n - 1), \tag{1.42}$$

$$\beta = (1 - p)(n - 1), \tag{1.43}$$

where θ has been estimated by the sample proportion $\hat{\theta} = p = \sum_{i=1}^{n} x_i / n$. One can immediately verify that whenever the hyperparameters α and β are elicited as in (1.42) and (1.43), then $\alpha + \beta + 1 = n$. The elicited parameters reflect the amount of information provided by a sample of size n.

Some further practical examples will be provided throughout the book. For an extended discussion of prior elicitation, the reader can refer to Garthwaite et al. (2005) and O'Hagan et al. (2006).

1.11   Sensitivity Analysis

In Sect. 1.4, it has been emphasized that the Bayes factor is not a measure of the relative support for the competing propositions provided by the data alone. The Bayes factor is influenced by the choice and the elicitation of the subjective prior densities (probabilities) for model parameters under propositions H_1 and H_2. This reflects background knowledge that may be available to analysts. For this reason, prior elicitation of model parameters must not be confused with prior probabilities of the propositions of interest.

While the computation of the Bayes factor requires prior assessments about unknown quantities, a main objection to the choice of such prior distributions is that they may be hard to define, in particular when the available information is limited. Situations characterized by an abundance of relevant data that can be used to construct a prior distribution may be rare. Generally, the choice of a prior is the result of a subtle combination of relevant information, published data, and explainable personal knowledge of the expert. The specification of the prior must be taken seriously, because it can be shown that even when a large amount of evidence is available, the marginal likelihood is highly sensitive to the choice of the prior distribution, and so is the Bayes factor (Gelman et al., 2014). This is different for the posterior distribution that is dominated by the likelihood.

Sensitivity analyses allow one to explore how results may be affected by changes in the priors (e.g. Kass & Raftery, 1995; Kass, 1993; Liu & Aitkin, 2008). This, however, may turn out to be computationally intensive and time consuming. An alternative approach has been proposed by Sinharay and Stern (2002) for comparing nested models, though it can be extended to non-nested models. The general idea is to assess the sensitivity of the Bayes factor to the prior distribution for a given parameter θ by computing the Bayes factor for a vector of parameter values (or a grid of parameter values in the case of a two-dimensional vector parameter θ). The result is a graphical representation of the Bayes factor (i.e., a sensitivity curve) as a function of θ, say BF_θ. In this way, one can get an idea about the Bayes factor one could obtain for different values of θ, and thus about the sensitivity of the Bayes factor to various prior distributions. These prior distributions have their mass concentrated on different apportionments of the parameter space. For one or two-dimensional problems, the inspection of a sensitivity curve represents a straightforward and effective approach to study the impact of varying parameter values on the BF under consideration. An example is given in Sect. 2.3.1 for the choice of the prior distribution about a Normal mean. A sensitivity analysis with respect to the prior probability assessments of competing propositions is provided in Sect. 3.2.3.

A further layer of sensitivity analyses relates to the choice of the utility/loss function. An example is presented in Sect. 2.2.3 for the choice of the loss function in the context of inference and decision about a population proportion. Section 4.4.1.2 gives an example for the investigation of the effect of different prior probabilities and loss values in the context of classification of skeletal remains.

A sensitivity analysis for Monte Carlo and Markov chain Monte Carlo procedures is presented in Sects. 2.2.2 and 3.4.1.3. In Sect. 4.3.3, a sensitivity analysis is developed for the choice of a smoothing parameter in a kernel density estimation.

1.12 Using R

R is a rich environment for data analysis and statistical computing. In its base package, it contains a large collection of functions for exploring, summarizing, and representing data graphically, handling many standard probability distributions and more. R includes a simple programming language that users can extend with new functions. Some basic instructions on the use of R or of particular functions are available from the R Help menu, by using the command `help.start()`. The reader can refer to, for example, Verzani (2014) for a detailed introduction to the use of R for descriptive and inferential statistics, to Albert (2009) for an overview of elements of Bayesian computation with R, and to the R project home page (https://www.r-project.org) for more references. Datasets and routines used in the examples throughout this book are available on the website of this book (on http://link.springer.com/).

Generally, we will give results of R computations as produced directly by R. We do not make any recommendations as to the level of precision that scientists should use when reporting numerical results.

Published with the support of the Swiss National Science Foundation (Grant no. 10BP12_208532/1).

Chapter 2
Bayes Factor for Model Choice

2.1 Introduction

The Bayes factor can assist forensic scientists in the evaluation of findings when recipients of expert information need help in discriminating between propositions concerning, for example, a parameter of interest. A typical example is the discrimination between competing propositions regarding the concentration of a controlled substance (e.g., drugs in blood) with respect to a given threshold. This chapter will approach one-sided hypothesis testing involving model parameters in the form of a proportion (Sect. 2.2) and a mean (Sect. 2.3). In both situations, additional factors, such as errors (Sects. 2.2.2 and 2.3.2), are considered. Aspects of decision-making are also considered (Sects. 2.2.3 and 2.3.3).

Throughout this chapter, the Bayes factor will be obtained as a ratio of marginal likelihoods following the ideas described in Sect. 1.4. The greater marginal likelihood will support the respective proposition against the other. This, along with other aspects, such as the decision maker's preferences among adverse consequences, has an impact on the decision-making process.

2.2 Proportion

A common problem in forensic practice is the investigation of the proportion of items or individuals that present a characteristic of interest, e.g., the proportion of seized pills containing a controlled substance or the proportion of counterfeit medicines in a given population. A consignment of items is considered a random sample from a super-population of items of the same type, and the parameter θ is the proportion of units in the super-population that present the target characteristic. Note that for consignments of very large size (i.e., several thousands), a finite number of units will correspond to each positive value of θ. For consignments of small size

S. Bozza et al., *Bayes Factors for Forensic Decision Analyses with R*,
Springer Texts in Statistics, https://doi.org/10.1007/978-3-031-09839-0_2

(i.e., smaller than 50), the parameter θ is a nuisance parameter (i.e., one that is not of primary interest) that can be integrated out, leaving a probability distribution for the unknown number of items having the target characteristic. For consignments of intermediate size, θ can be treated as a continuous value in the interval $(0, 1)$ (e.g., Aitken et al., 2021). As an example, consider the following pair of propositions:

H_1:   The proportion θ of items having the characteristic of interest is larger than
        θ_0.
H_2:   The proportion θ of items having the characteristic of interest is smaller than
        or equal to θ_0,

where $\theta_0 \in (0, 1)$ is a given threshold of legal interest.[1] Note that applications of this type of propositions are broad and include, for example, quality control of food (and other consumer products), the analysis of levels of contamination of laboratory equipment, and the extent of environmental pollution.

This section covers three main topics: (1) inference about an unknown proportion θ (Sect. 2.2.1), (2) inference about θ when background elements may affect the counting process (Sect. 2.2.2), and (3) decision regarding competing propositions about θ (Sect. 2.2.3).

2.2.1   Inference About a Proportion

Consider a case of inference about a population parameter based on a sample of size n. Aitken (1999) and Aitken et al. (2021) discuss how to choose a sample size. Suppose that among the n items, x shows a characteristic that is of interest from a legal point of view. The question then is how such an analytical result supports one or the other of the competing propositions regarding the proportion of items in the population that have the target characteristic.

Experiments of this kind can be regarded as Bernoulli trials (after the Swiss mathematician Jacob Bernoulli, 1654–1705), where trials are independent and give rise to one of the two mutually exclusive outcomes, conventionally labeled success and failure, with constant probability of success in each trial. The binomial distribution $\mathrm{Bin}(n, \theta)$ is a statistical model for data that arise from a sequence of Bernoulli trials:

$$f(x \mid n, \theta) = \binom{n}{x} \theta^x (1 - \theta)^{n-x}, \qquad\qquad x = 0, 1, \ldots, n.$$

In the Bayesian perspective, the most common prior distribution for the parameter of interest θ is the beta distribution $\mathrm{Be}(\alpha, \beta)$:

---

[1] See Biedermann et al. (2012, 2018) for a general discussion of thresholds of legal interest when data are continuous.

$$f(\theta \mid \alpha, \beta) = \theta^{\alpha-1}(1-\theta)^{\beta-1}/\mathrm{B}(\alpha, \beta), \qquad\qquad 0 < \theta < 1 \; ; \; \alpha, \beta > 0,$$

with $\mathrm{B}(\alpha, \beta) = \frac{\Gamma(\alpha)\Gamma(\beta)}{\Gamma(\alpha+\beta)}$.

The beta prior distribution and the binomial likelihood are conjugate (see Sect. 1.10): after inspecting a sample, one can easily compute the posterior distribution, which is still beta, $\mathrm{Be}(\alpha^*, \beta^*)$ with parameters updated according to well-known updating rules, $\alpha^* = \alpha + x$, $\beta^* = \beta + n - x$ (e.g., Lee, 2012). The prior odds, the posterior odds, and the Bayes factor can be easily computed, as discussed in Sect. 1.4, by means of standard routines.

Example 2.1 (Counterfeit Medicines) Consider a case in which a large batch of medicines (say, $N > 50$) is seized, suspected to contain counterfeit items. The following propositions are of interest:

H_1: The proportion θ of counterfeit medicines is greater than 0.2.
H_2: The proportion θ of counterfeit medicines in not greater than 0.2.

Suppose that, initially, limited information is available so that a uniform prior distribution is chosen over the interval $(0, 1)$, that is, $\theta \sim \mathrm{Be}(1, 1)$. Note that although a prior distribution $\mathrm{Be}(1, 1)$ is often called *uninformative*, it is in fact informative (see Sect. 1.10 and de Finetti (1993b)). It conveys the view that every value of θ in the interval $(0, 1)$ is considered equally probable. The prior odds can then easily be obtained.

```
> th=0.2
> a=1
> b=1
> pi1=pbeta(th,a,b,lower.tail=F)
> prior_odds=pi1/(1-pi1)
> prior_odds

[1] 4
```

A uniform prior distribution clearly favors, a priori, hypothesis H_1, that θ is greater than 0.2. Next, suppose that a sample of size 40 is analyzed and 12 out 40 items are found to be positive (counterfeit). The posterior distribution follows immediately and so the posterior odds and the Bayes factor.

```
> n=40
> x=12
> astar=a+x
> bstar=b+n-x
> alpha1=pbeta(th,astar,bstar,lower.tail=F)
```

(continued)

Example 2.1 (continued)
```
> post_odds=alpha1/(1-alpha1)
> post_odds
```

[1] 18.19594

The posterior probability of proposition H_1 is, therefore, approximately 18 times greater than the posterior probability of the alternative proposition H_2. Thus, the Bayes factor can be obtained as

```
> BF=post_odds/prior_odds
> BF
```

[1] 4.548985

The Bayes factor indicates that the evidence is in favor of proposition H_1 that the proportion of counterfeit medicines is greater than 0.2, rather than proposition H_2 (i.e., $\theta < 0.2$). According to the verbal scale presented in Table 1.2, the BF weakly supports proposition H_1 over H_2.

To help specify the prior distribution, information in the form of data regarding similar consignments from cases with comparable circumstances may be used. Such data may suggest a distribution other than the uniform distribution used in the above example. An example of how to elicit a subjective prior distribution about a proportion is provided in Sect. 1.10. For a more extensive discussion about prior elicitation for a proportion, the reader can refer to O'Hagan et al. (2006). Forensically relevant applications of prior elicitation for θ are discussed in Aitken (1999). Note, however, that in certain practical applications, analytical results may be affected by further factors that cannot be dissociated from the observational process. An example for such a factor is considered is Sect. 2.2.2.

The analysis pursued above focused on the problem of inference about a proportion for a large batch. Consider now the case where the size N of the consignment is small (less than 50). Suppose a sample of size n is inspected and x items are found to present the target characteristic (e.g., yield a positive test result), so that $\theta \sim \text{Be}(\alpha + x, \beta + n - x)$. Denote by Y the unknown number of positive items in the uninspected part of the consignment. This random variable has still a binomial distribution, $Y \sim \text{Bin}(m, \theta)$, where $m = N - n$ represents the number of units that have not been inspected. The probability distribution for the unknown number of positive units can be obtained by integrating out parameter θ. This turns out to be a beta-binomial distribution $\text{Be-Bin}(n, m, x, \alpha, \beta)$:

$$\Pr(Y = y \mid n, m, x, \alpha, \beta)$$

$$= \frac{\Gamma(n + \alpha + \beta)\binom{m}{y}\Gamma(y + x + \alpha)\Gamma(n + m - x - y + \beta)}{\Gamma(x + \alpha)\Gamma(n - x + \beta)\Gamma(n + m + \alpha + \beta)}(y = 0, 1, \ldots, n)$$

$$(2.1)$$

(Aitken, 1999).

Example 2.2 (Counterfeit Medicines—Small Consignment) Consider Example 2.1 and suppose now that the consignment is small, say $N = 40$. Suppose further that a sample of size $n = 10$ has been inspected and that 2 items are found to be counterfeit. Starting from a uniform prior distribution $\theta \sim \text{Be}(1, 1)$, the beta posterior distribution becomes $\theta \sim \text{Be}(3, 9)$.

```
> N=40
> n=10
> x=2
> a=1
> b=1
> astar=a+x
> bstar=b+n-x
```

The distribution of Y then is Be-Bin(10, 30, 2, 1, 1). The probability to observe a given number of counterfeit items (e.g., $y = 1$) in the remainder of the consignment can be obtained using the function dbbinom that is available in the package extraDistr (Wolodzko, 2020).

```
> library(extraDistr)
> dbbinom(1,N-n,astar,bstar)

[1] 0.03665943
```

One can also use the function pbbinom that allows to compute the cumulative distribution function for the beta-binomial random variable in (2.1). For example, the probability to observe at most 2 counterfeit items can be obtained as

```
> pbbinom(2,N-n,astar,bstar)

[1] 0.109604
```

A Bayesian network for inference about a proportion of a small consignment has been developed in Biedermann et al. (2008). Posterior probabilities for θ can easily be calculated with such models.

2.2.2   Background Elements Affecting Counting Processes

In many real-world applications, counting processes performed in forensic laboratories cannot be considered error-free. Examinations may be affected by inefficiencies and perturbing factors. For example, it may be that items are lost or missed during counting or that background elements are present, i.e., objects observationally indistinguishable from the target objects. This section addresses inferential challenges due to such background elements.

Suppose that x is the number of recorded successes, i.e., the number of times that the target characteristic is detected. However, the number x may not correspond to the number x_s of items actually showing the characteristic of interest but be affected by a certain number of background elements, x_b, that are wrongly counted as successes. This complication may typically arise in applications where the items of interest are small particles. Consider, for example, the assessment of rice quality in a context of food quality control. Rice quality can be measured by means of several features, such as the percentage of cracked or immature grains. For example, there may be legal provisions regarding the maximum tolerated amount of cracked grains.[2] It might then be of interest to compare alternative propositions according to which the percentage of cracked grains is above or below a given regulatory threshold. A key question is how to conduct such a comparison when the counting process may be affected by background elements, e.g., oil seeds in the example here.

While the number of elements *actually* showing the target characteristic is modeled as the outcome of a binomial distribution, $X_s \sim \text{Bin}(n, \theta)$, the amount of background elements affecting the counting process, x_b, can be modeled by a Poisson distribution, $X_b \sim \text{Pn}(\lambda)$, where λ is the mean number of background elements (D'Agostini, 2004). The total number of *recorded* successes is therefore $X = X_s + X_b$. The graphical model (see e.g. Cowell et al., 1999) in Fig. 2.1 offers a schematic representation of the probabilistic relationship among the variables.

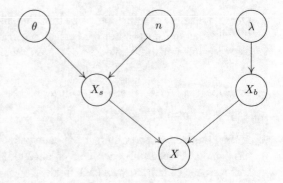

Fig. 2.1 Graphical representation of the statistical model for inference about a population proportion based on count data in presence of background elements affecting counting processes

---

[2] For legislation in, e.g., Italy, see Gazzetta Ufficiale della Repubblica Italiana, 6, 09-01-2018, Decreto 20 settembre 2017.

It can be shown[3] that X has the following probability distribution:

$$f(x \mid n, \theta, \lambda) = \sum_{x_b=0}^{x} \binom{n}{x - x_b} \theta^{x-x_b} (1 - \theta)^{n-x+x_b} e^{-\lambda} \lambda^{x_b} / x_b!$$

Recall that prior uncertainty about θ can modeled by a beta distribution $\text{Be}(\alpha, \beta)$. The posterior distribution is then given by

$$f(\theta \mid n, x, \lambda) = \frac{\sum_{x_b=0}^{x} \binom{n}{x-x_b} \theta^{x-x_b} (1 - \theta)^{n-x+x_b} e^{-\lambda} \lambda^{x_b} / x_b! \theta^{\alpha-1} (1 - \theta)^{\beta-1}}{f(x \mid n, \lambda) B(\alpha, \beta)}, \quad (2.2)$$

where the normalizing constant $f(x \mid n, \lambda)$ in the denominator is

$$f(x \mid n, \lambda) = \int f(x \mid n, \theta, \lambda) f(\theta) d\theta. \quad (2.3)$$

The posterior distribution (2.2) cannot be obtained in closed form as the integral characterizing the normalizing constant $f(x \mid, n, \lambda)$ is not tractable analytically. However, since it is possible to draw values from the beta distribution, the integral in (2.3) can be computed by Monte Carlo approximation as in (1.30), that is,

$$\hat{f}(x \mid n, \lambda) = \frac{1}{N} \sum_{i=1}^{N} f(x \mid n, \theta^{(i)}, \lambda), \quad (2.4)$$

where $\theta^{(i)} \sim \text{Be}(\alpha, \beta)$.

Example 2.3 (Rice Quality) Consider a consignment of rice and suppose that it is of interest to assess whether the proportion of cracked grains is below a given level of tolerance. The following competing propositions may be of interest:

H_1: The proportion θ of cracked grains is greater than 0.025.
H_2: The proportion θ of cracked grains is smaller than or equal to 0.025.

In a sample of 1000 grains, a total of 28 cracked grains are observed.

(continued)

---

[3] The method for finding the distribution of a sum of random variables is given, for example, in Casella and Berger (2002). It can be used to extend the model to the case of missing counts, an aspect that is not treated here.

Example 2.3 (continued)
```
> n=1000
> x=28
```

The beta prior distribution for θ needs to be elicited. Suppose that available knowledge indicates that it is implausible that the proportion of cracked grains is greater than 5%. An asymmetric prior distribution with a prior mass concentrated over values lower than 0.05 can be elicited as follows. Start with $\alpha = 1$ and $\beta = 1$, then increment β by 1 until the shape of the beta distribution is such that $\Pr(\theta > 0.05)$ is small, e.g., equal to 0.1.

```
> a=1
> b=1
> while(pbeta(0.05,a,b,lower.tail=F)>0.1){b=b+1}
> c(a,b,pbeta(0.05,a,b,lower.tail=F))

[1]   1.00000000 45.00000000   0.09944026
```

The parameters α and β can thus be set equal to 1 and 45, respectively. Figure 2.2 (left) can be obtained with

```
> plot(function(x) dbeta(x,a,b),0,0.1,xlab=expression
+ (theta),ylab=expression(paste(pi)*paste('(')*
+ paste(theta)*paste(')')))
```

The prior odds can now be computed in a straightforward manner.

```
> th0=0.025
> pi1=pbeta(th0,a,b,lower.tail=F)
> prior_odds=pi1/(1-pi1)
> prior_odds

[1]  0.4706802
```

This value, approximately 0.5, means that the probability of hypothesis H_2 is, a priori, approximately 2 times greater than the probability of hypothesis H_1.

Suppose that when inspecting a sample of 1000 rice grains, on average, 1 grain (e.g., oil seed) is wrongly counted as cracked. Parameter λ can thus be taken to be equal to 0.001.

First, we write a function dbinpois that computes the product between a binomial likelihood $Bin(n, \theta)$ at $x - x_b$ and a Poisson likelihood $Pn(\lambda)$ at x_b.

```
> dbinpois=function(xb){
+ dbinom((x-xb),n,theta)*dpois(xb,lambda)}
```

The unnormalized posterior distribution in (2.2)

(continued)

Example 2.3 (continued)

$$\frac{\sum_{x_b=0}^{x} \binom{n}{x-x_b} \theta^{x-x_b}(1-\theta)^{n-x+x_b} e^{-\lambda} \lambda^{x_b}/x_b! \theta^{\alpha-1}(1-\theta)^{\beta-1}}{B(\alpha, \beta)}$$

is computed as

```
> lambda=0.001
> xb=matrix(seq(0,x,1),nrow=1)
> incr=0.0001
> thetav=seq(0.0001,0.9999,incr)
> theta=thetav[1]
> s=sum(apply(xb,2,dbinpois))
> upost=dbeta(theta,a,b)*s
> for (i in 2:length(thetav)){
+          theta=thetav[i]
+          s=sum(apply(xb,2,dbinpois))
+          upost=c(upost,dbeta(theta,a,b)*s)
+          }
```

The normalizing constant $f(x \mid n, \lambda)$ can be approximated as in (2.4)

```
> theta=rbeta(1,a,b)
> norm_const=sum(apply(xb,2,dbinpois))
> nn=10000
> for (i in 2:nn){
+          theta=rbeta(1,a,b)
+          s=sum(apply(xb,2,dbinpois))
+          norm_const=norm_const+s
+          }
> norm_const=norm_const/nn
```

and the approximated posterior density, represented in Fig. 2.2 (right), can be
obtained as

```
> normpost=upost/(norm_const)
> plot(thetav,normpost,xlab=expression(paste(theta)),
+ ylab=expression(hat(f)*paste('(')*paste(theta)*
+ paste('|n,x,')*paste(lambda)*paste(')')),
+ xlim=c(0,0.1),type='l')
```

To calculate the BF, we need to obtain the posterior probabilities of the competing
propositions H_1 and H_2. Consider proposition H_2. The (approximate) posterior
probability of proposition H_2 can be obtained by Monte Carlo integration as

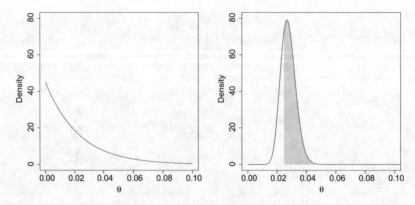

Fig. 2.2 Left: beta prior distribution $Be(1, 45)$ of the unknown proportion θ of cracked grains (Example 2.3). Right: approximated posterior distribution of θ, $\hat{f}(\theta \mid n, x, \lambda)$. The gray shaded area shows the posterior probability of the hypothesis H_1 ($\theta > 0.025$)

$$
\begin{aligned}
\hat{\alpha}_2 &= \frac{1}{\hat{f}(x \mid n, \lambda)} \int_0^{\theta_0} f(x \mid n, \theta, \lambda) f(\theta) d\theta \\
&= \frac{\theta_0}{\hat{f}(x \mid n, \lambda)} \int_0^{\theta_0} f(x \mid n, \theta, \lambda) f(\theta) \frac{1}{\theta_0} d\theta \\
&\approx \frac{\theta_0}{\hat{f}(x \mid n, \lambda)} \cdot \frac{1}{N} \sum_{i=1}^{N} f(x \mid n, \theta^i, \lambda) f(\theta^i) d\theta,
\end{aligned} \tag{2.5}
$$

where θ^i is sampled from a uniform distribution in the interval $(0, \theta_0)$, $\theta^i \sim$ Unif$(0, \theta_0)$, and the normalizing constant $f(x \mid n, \lambda)$ is approximated as in (2.4). The (approximate) posterior probability of hypothesis H_1 is $1 - \hat{\alpha}_2$. The (approximated) BF will be

$$
\widehat{BF} = \frac{\widehat{\alpha_1}/\widehat{\alpha_2}}{\pi_1/\pi_2}. \tag{2.6}
$$

Example 2.4 (Rice Quality—Continued) Consider the scenario described in Example 2.3, and compute the (approximate) posterior probability of the hypothesis H_2: the proportion θ of cracked grains is smaller than or equal to 0.025 (as in 2.5).

(continued)

Example 2.4 (continued)

```
> m=10000
> theta=runif(m,0,th0)
> alpha2=mean(rowSums(apply(xb,2,dbinpois))
+ *dbeta(theta,a,b))*th0/norm_const
> alpha2

[1] 0.30753
```

The (approximate) posterior probability of hypothesis H_1 then is $\hat{\alpha}_1 = 0.6925$. This is highlighted by the gray shaded area in Fig. 2.2 (right). The posterior odds and the BF therefore are

```
> post_odds=(1-alpha2)/(alpha2)
> post_odds

[1] 2.251715

> BF=post_odds/prior_odds
> BF

[1] 4.783959
```

The Bayes factor indicates that the evidence favors hypothesis H_1, i.e., $\theta > 0.025$, over H_2, i.e., $\theta \leq 0.025$. A BF of approximately 5 provides limited support for the hypothesis H_1. Note that the results obtained by the laboratory analyses clearly affect our belief about θ. The analytical results change prior odds in favor of H_1 (0.47) to posterior odds of approximately 2.25 in favor of H_1.

2.2.2.1 Sensitivity to Monte Carlo Approximation

The Monte Carlo estimate of the Bayes factor obtained in (2.6) is subject to variability, which may be a source of concern. Figure 2.3 provides an illustration of BF variability. The figure shows 500 realizations of the BF approximation in (2.6).

```
> ns=500
> m=10000
> BFs=0
> dbinpois=function(xb){
+ dbinom((x-xb),n,theta)*dpois(xb,lambda)}
> for (j in 1:ns){
+             rthetav=rbeta(m,a,b)
```

```
+           norm_const=0
+           for (i in 1:m){
+                   theta=rthetav[i]
+                   s=sum(apply(xb,2,dbinpois))
+                   norm_const=norm_const+s
+           }
+           norm_const=norm_const/m
+           theta=runif(m,0,th0)
+           alpha2=mean(rowSums(apply(xb,2,dbinpois))
+           *dbeta(theta,a,b))*th0/norm_const
+           post_odds=(1-alpha2)/alpha2
+           BFs=c(BFs,post_odds/prior_odds)
+ }
> BFs=BFs[-1]
> hist(BFs,main='',prob=T)
> curve(dnorm(x,mean(BFs),sd(BFs)),lwd=2,add=T)
```

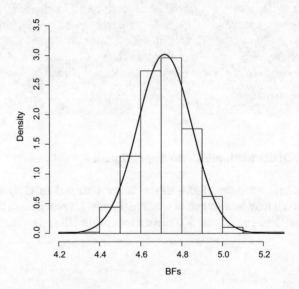

Fig. 2.3 Histogram of 500 realizations of the BF approximation in (2.6), where the posterior probability of hypothesis H_2 is obtained as in (2.5). The solid line represents the fitted Normal density

The purpose of the graphical representation in Fig. 2.3 is to illustrate that the repeated application of the procedure leads to a distribution of BFs. While the Monte

Carlo estimate is not an exact value, it can be shown that the approximation error can be made arbitrarily small by generating a sufficiently large amount of observations. For a large number of simulations, it can also be proven, by Central Limit Theorem, that the error $| \hat{f}(x) - f(x) | \sqrt{N}$ is normally distributed. This can be used to analyze the variability of the Monte Carlo estimate (see, e.g., Marin and Robert (2014)). Note that the shape of the histogram is roughly symmetric and bell-shaped, as shown in Fig. 2.3.

It is worth noting that other, more efficient ways than traditional Monte Carlo methods may be implemented to compute the integrals related to the posterior probabilities of the competing hypotheses. Importance sampling (see Sect. 1.8), for example, can improve the integral approximation. It can also be used when the target density is unnormalized. Consider again the posterior probability of hypothesis H_2:

$$\alpha_2 = \int_0^{\theta_0} \frac{f(x \mid n, \theta, \lambda) f(\theta)}{f(x \mid n, \lambda)} d\theta.$$

This can be rewritten as

$$\alpha_2 = \frac{1}{f(x \mid n, \lambda)} \int_0^1 h(\theta) f(x \mid n, \theta, \lambda) f(\theta) \frac{g(\theta)}{g(\theta)} d\theta$$

$$= \frac{1}{f(x \mid n, \lambda)} \int_0^1 h(\theta) w(\theta) g(\theta) d\theta,$$

where

$$h(\theta) = \begin{cases} 1 \text{ if } 0 < \theta < \theta_0 \\ \\ 0 \text{ if } \theta_0 \leq \theta < 1, \end{cases}$$

$w(\theta) = f(x \mid n, \theta, \lambda) f(\theta) / g(\theta)$ and $g(\theta)$ is the importance sampling function.

The posterior probability α_2 can be approximated as

$$\hat{\alpha}_2 = \frac{\frac{1}{N} \sum_{i=1}^N h(\theta^i) w(\theta^i)}{\frac{1}{N} \sum_{i=1}^N w(\theta^i)}, \tag{2.7}$$

where $\theta^i \sim g(\theta)$.

Example 2.5 (Rice Quality—Continued) A Be(20, 780) is chosen as importance sampling function $g(\theta)$. It can be readily verified that it is centered at 0.025 and that the density rapidly collapses toward zero for values greater than 0.04. This will avoid the generation of points for which the integrand is close to zero, with a very modest contribution to the approximation. Next, sample 10000 values from this distribution.

```
> m=10000
> a1=20
> b1=780
> theta=rbeta(m,a1,b1)
```

The posterior probability α_2 of hypothesis H_2 can be obtained as in (2.7)

```
> fx=rep(0,m)
> fx[theta<th0]=1
> num=mean(rowSums(apply(xb,2,dbinpois))*
+ dbeta(theta,a,b)/dbeta(theta,a1,b1)*fx)
> den=mean(rowSums(apply(xb,2,dbinpois))*
+ dbeta(theta,a,b)/dbeta(theta,a1,b1))
> alpha2=num/den
> alpha2

[1] 0.3079344

> BF=((1-alpha2)/alpha2)/prior_odds
> BF

[1] 4.774886
```

Figure 2.4 provides an illustration of BF variability. Notice that while the BFs in Figs. 2.3 and 2.4 have roughly the same location, the importance sampling in (2.7) produced an increase in precision.

It is important to understand that the resulting distribution does *not* mean that there is a distribution *for a given* BF because the BF, by definition, is a single number. See, e.g., Taroni et al. (2016) and Biedermann et al. (2017a) for discussions of this topic among forensic statisticians and forensic scientists. The error resulting from the implementation of numerical techniques is an important source of information about which the scientist should be transparent. Following ideas presented in Tanner (1996), recently reconsidered by Ommen et al. (2017) in a forensic context, the numerical precision in the overall approximated value can be estimated by the associated Monte Carlo standard error.

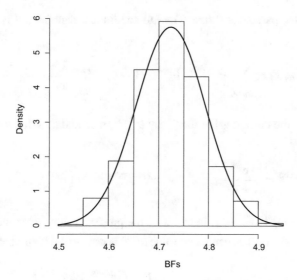

Fig. 2.4 Histogram of 500 realizations of the BF approximation in (2.6), where the posterior probability of hypothesis H_2 is obtained as in (2.7). The solid line represents the fitted Normal density

2.2.2.2 Unknown Expected Value of the Number of Background Elements

It is important to note that, contrary to what was developed in Example 2.3, the expected value λ of the number of background events is generally unknown. The uncertainty about λ can be modeled by means of a gamma distribution, $\lambda \sim$ Ga(a, b). The marginal posterior distribution of parameter θ, written $f(\theta \mid n, x)$, now takes a more complicated form as one needs to handle the joint posterior distribution that is proportional to

$$f(\theta, \lambda \mid n, x)$$
$$\propto \sum_{x_b=0}^{x} \binom{n}{x - x_b} \theta^{x-x_b} (1 - \theta)^{n-x+x_b} \frac{e^{-\lambda} \lambda^{x_b}}{x_b!} \theta^{\alpha-1} (1 - \theta)^{\beta-1} \lambda^{a-1} e^{-b\lambda}.$$

$$(2.8)$$

Following ideas described in Taroni et al. (2010), a two-block M–H algorithm (Sect. 1.8) can be implemented in order to draw a sample from the joint posterior distribution in (2.8). For each block, the candidate generating density is taken to be Normal with the mean equal to the current value of the parameter and the variance chosen so as to obtain a good acceptance rate (Gamerman & Lopes, 2006).

Consider the parameter θ first. The full conditional density of θ is proportional to

$$f_1(\theta \mid \lambda, n, x) \propto \sum_{x_b=0}^{x} \binom{n}{x - x_b} \theta^{x-x_b}(1 - \theta)^{n-x+x_b} \frac{\lambda^{x_b}}{x_b!} \theta^{\alpha-1}(1 - \theta)^{\beta-1}.$$

Starting from the current value for θ, say $\theta^{(i-1)}$, a candidate value θ^{prop} for θ can be obtained as

$$\theta^{\text{prop}} = \frac{e^{\psi^{\text{prop}}}}{1 + e^{\psi^{\text{prop}}}}, \qquad \text{where } \psi^{\text{prop}} \sim N\left(\psi^{(i-1)}, \tau_1^2\right),$$

and $\psi^{(i-1)} = \log\left(\frac{\theta^{(i-1)}}{1-\theta^{(i-1)}}\right)$. In this way, the proposed value θ^{prop} will be defined in the interval $(0, 1)$. The candidate value θ^{prop} is accepted with probability

$$\alpha(\psi^{(i-1)}, \psi^{\text{prop}}) = \min\left\{1, \frac{f(\psi^{\text{prop}} \mid \lambda^{(i-1)})}{f(\psi^{(i-1)} \mid \lambda^{(i)})}\right\},$$

where $f(\psi \mid \lambda)$ is the reparametrized full conditional density of parameter θ and can be obtained as

$$f(\psi \mid \lambda) = \frac{e^{\psi}}{(1 + e^{\psi})^2} f_1\left(\frac{e^{\psi}}{(1 + e^{\psi})^2} \mid \lambda, n, x\right).$$

See, e.g., Casella and Berger (2002) for distributions of functions of random variables.

If the candidate θ^{prop} is accepted, it becomes the current value of the chain, i.e., $\theta^{(i)} = \theta^{\text{prop}}$; otherwise $\theta^{(i)} = \theta^{(i-1)}$.

The second block refers to parameter λ. The full conditional density of parameter λ is proportional to

$$f_2(\lambda \mid \theta, n, x) \propto \sum_{x_b=0}^{x} \binom{n}{x - x_b} \theta^{x-x_b}(1 - \theta)^{n-x+x_b} \frac{e^{-\lambda}\lambda^{x_b}}{x_b!} \lambda^{a-1} e^{-b\lambda}.$$

Starting from the current value for λ, say $\lambda^{(i-1)}$, a candidate value λ^{prop} for λ can be obtained as

$$\lambda^{\text{prop}} = e^{\phi^{\text{prop}}}, \qquad \text{where } \phi^{\text{prop}} \sim N\left(\phi^{(i-1)}, \tau_2^2\right),$$

and $\phi^{(i-1)} = \log \lambda^{(i-1)}$. In this way, the proposed value λ^{prop} will be defined in the interval $(0, \infty)$. The candidate value λ^{prop} is accepted with probability

$$\alpha(\phi^{(i-1)}, \phi^{\text{prop}}) = \min\left\{1, \frac{f(\phi^{\text{prop}} \mid \theta^{(i-1)})}{f(\phi^{(i-1)} \mid \theta^{(i-1)})}\right\},$$

where $f(\phi \mid \theta)$ is the reparametrized full conditional density of parameter λ and can be obtained as

$$f(\phi \mid \theta) = e^{\phi} f_2(e^{\phi} \mid \theta, n, x).$$

If the candidate λ^{prop} is accepted, it becomes the current value of the chain, i.e., $\lambda^{(i)} = \lambda^{\text{prop}}$; otherwise $\lambda^{(i)} = \lambda^{(i-1)}$.

The two-block M–H algorithm can be summarized as follows:
Initialization: start with arbitrary values $\theta^{(0)}$ and $\lambda^{(0)}$
Iteration i:

1. Given $\theta^{(i-1)}$ and $\lambda^{(i-1)}$,

   – Generate θ^{prop} according to $f_1(\theta \mid \lambda^{(i-1)}, n, x)$.
   – With probability $\alpha(\theta^{(i-1)}, \theta^{\text{prop}})$ accept θ^{prop} and set $\theta^{(i)} = \theta^{\text{prop}}$; otherwise reject θ^{prop} and set $\theta^{(i)} = \theta^{(i-1)}$.

2. Given $\theta^{(i)}$ and $\lambda^{(i-1)}$,

   – Generate λ^{prop} according to $f_2(\lambda \mid \theta^{(i)}, n, x)$.
   – With probability $\alpha(\lambda^{(i-1)}, \lambda^{\text{prop}})$ accept λ^{prop} and set $\lambda^{(i)} = \lambda^{\text{prop}}$; otherwise reject λ^{prop} and set $\lambda^{(i)} = \lambda^{(i-1)}$.

Return $\{\theta^{(n_b+1)}, \dots, \theta^{(N)}\}$ and $\{\lambda^{(n_b+1)}, \dots, \lambda^{(N)}\}$,
where n_b is the burn-in period and N is the number of iterations.

Example 2.6 (Rice Quality—Continued) Consider again Example 2.3 where prior uncertainty about θ was modeled by a Be(1, 45) distribution, and the parameter λ was set equal to 0.001. For the purpose of the example here, a gamma distribution with parameters $a = 2$ and $b = 1000$ is used to model prior uncertainty about λ. The prior density Ga(2, 1000) is shown in Fig. 2.5. It can be observed that the prior mass is concentrated at very small values of λ.

```
> n=1000
> x=28
```

(continued)

Fig. 2.5 Gamma prior
distribution Ga(2, 1000) over
λ for $\lambda \in (0, 0.01)$

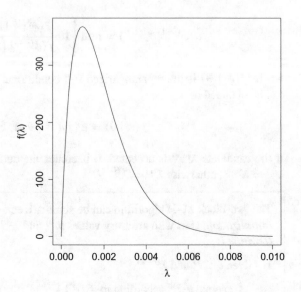

Example 2.6 (continued)
```
> xb=matrix(seq(0,x,1),nrow=1)
> a=1
> b=45
> ag=2
> bg=1000
> plot(function(x) dgamma(x,2,1000),0,0.01,xlab=
+ expression(lambda),ylab=expression(paste('f(')*
+ paste(lambda)*paste(')')))
```

Let the starting values for θ and λ be $\theta^{(0)} = 0.1$ and $\lambda^{(0)} = 0.001$, and the variances τ_1^2 and τ_2^2 of the proposal densities be set equal to 0.7 and 3, respectively.

```
> theta=0.1
> lambda=0.001
> tau=c(0.7,3)
```

Current values of the parameters θ and λ will be stored in a vector called
`thetav` and `lambdav`, respectively.

```
> thetav=theta
> lambdav=lambda
```

(continued)

Example 2.6 (continued)

Before running the algorithm, it is useful to introduce the following functions: mh1 is used to obtain the candidate (current) value θ^{prop} (θ^{curr}); mh2 is used to calculate the probability of acceptance of the candidate value θ^{prop}; dbinpois computes the product between a binomial likelihood $\text{Bin}(n, \theta)$ at $x - x_b$ and a Poisson likelihood at x_b.

```
> mh1=function(x){x/(1+x)}
> mh2=function(x){x/((1+x)^2)}
> dbinpois=function(xb){
+ dbinom((x-xb),n,theta)*dpois(xb,lambda)}
```

The MCMC algorithm is run over 15000 iterations, with a burn-in range of 5000 iterations.

```
> n.iter=15000
> acct=n.iter
> accl=n.iter
> burn.in=5000
> for (i in 1:n.iter){
+ psicurr=log(theta/(1-theta))
+ s=sum(apply(xb,2,dbinpois))
+ pipsicurr=mh2(exp(psicurr))*dbeta(theta,a,b)*s
+
+ # Generate the candidate value of parameter theta
+
+ psiprop=rnorm(1,psicurr,tau[1])
+ theta=mh1(exp(psiprop))
+ s=sum(apply(xb,2,dbinpois))
+ pipsiprop=mh2(exp(psiprop))*dbeta(theta,a,b)*s
+
+ # acceptance/rejection of the candidate value
+ # (parameter theta)
+
+ if(runif(1)>pipsiprop/pipsicurr){
+ theta=mh1(exp(psicurr))
+ acct=acct-1}
+ thetav=c(thetav,theta)
+
+ # generate the candidate value of parameter lambda
+
+ phicurr=log(lambda)
+ s=sum(apply(xb,2,dbinpois))
```

(continued)

Example 2.6 (continued)

```
+ piphicurr=exp(phicurr)*dgamma(lambda,ag,bg)*s
+ phiprop=rnorm(1,phicurr,tau[2])
+ lambda=exp(phiprop)
+ s=sum(apply(xb,2,dbinpois))
+ piphiprop=exp(phiprop)*dgamma(lambda,ag,bg)*s
+
+ # acceptance/rejection of  the candidate value
+ # (parameter lambda)
+
+ if(runif(1)>piphiprop/piphicurr){
+ lambda=exp(phicurr)
+ accl=accl-1}
+ lambdav=c(lambdav,lambda)
+ }
> c(acct/n.iter,accl/n.iter)

[1] 0.3102000 0.2973333
```

These values represent the acceptance rates for θ and λ, respectively.

The output of the simulation run is shown in Fig. 2.6, representing the trace-plot, the autocorrelation plot (showing the correlation structure of the sequences), and the histogram of the simulated draws for θ (left column) and λ (right column). The simulated draws have an acceptance rate of approximately 31% for θ and 30% for λ. The trace-plots of simulated draws look like random noise and the autocorrelation decreases rapidly as the time lag at which it is calculated increases.

```
> par(mfrow=c(3,2))
> plot(thetav,type='l',xlab='Iterations',ylab=
+ expression(paste(theta)),main=expression(paste
+ (theta)))
> plot(lambdav,type='l',xlab='Iterations',ylab=
+ expression(paste(lambda)),main=expression(paste
+ (lambda)))
> acf(thetav[-c(1:burn.in)],type="correlation",ci=0,
+ main=expression(paste(theta)),ylab='')
> acf(lambdav[-c(1:burn.in)],type="correlation",ci=0,
+ main=expression(paste(lambda)),ylab='')
> hist(thetav[-c(1:burn.in)],xlab=expression(paste
+ (theta)),ylab='',main='')
> hist(lambdav[-c(1:burn.in)],xlab=expression(paste
+ (lambda)),ylab='',main='')
```

(continued)

Example 2.6 (continued)
Note that the argument `ci=0` in the function `acf` for computing and plotting the estimate of the autocorrelation function suppresses the plot of the confidence interval.

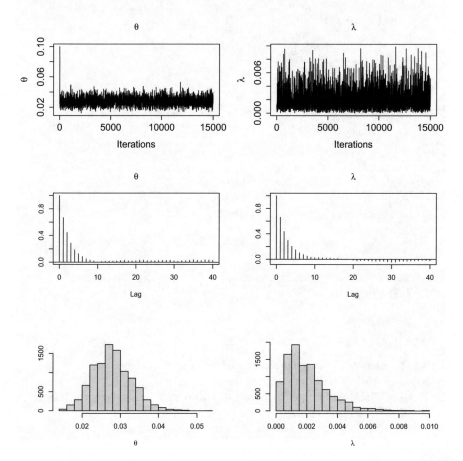

Fig. 2.6 MCMC diagnostic with trace-plots of simulated draws of θ (top left) and λ (top right), autocorrelation plots over the last 10000 iterations (center) and histograms over the last 10000 iterations (bottom)

The simulated values $\theta^{(n_b+1)}, \ldots, \theta^{(N)}$ can serve as draws from the posterior distribution $f_1(\theta \mid \lambda, n, x)$. The posterior probability of hypothesis H_1 can then be approximated as

$$\widehat{\alpha_1} = \sum_{\theta^{(i)} > 0.025} \theta^{(i)}/(N - n_b), \qquad (2.9)$$

and the BF can be obtained straightforwardly.

Example 2.7 (Rice Quality—Continued) Using a burn-in range of 5000 iterations, the average value of parameter θ over the last 10000 iterations can be computed as

```
> thetahat=mean(thetav[-c(1:burn.in)])
> thetahat

[1] 0.02788516
```

The posterior probability of hypothesis H_1 can be approximated as in (2.9):

```
> alpha1=sum(thetav[-c(1:burn.in)]>th0)/
+ (n.iter-burn.in)
> alpha1

[1] 0.71

> post_odds=alpha1/(1-alpha1)
> post_odds

[1] 2.448276
```

Recall that the prior odds have been quantified previously as approximately 0.47. The Bayes factor then is

```
> post_odds/prior_odds

[1] 5.201569
```

The uncertainty about the presence of background elements, modeled by λ, modifies the value of the BF from approximately 4.77 to 5.2. This change is small. The BF still provides only weak support for the hypothesis H_1 that $\theta > 0.025$, compared to H_2.

2.2.3   Decision for a Proportion

The normative framework for decision-making introduced in Chap. 1 is well suited for addressing problems of statistical inference presented in this chapter. Consider again a pair of competing propositions as defined in Sect. 2.2 regarding the question of whether the proportion of items showing a target characteristic of interest is

Fig. 2.7 Linear loss function
$L(d_1, \theta)$ (solid line) and
$L(d_2, \theta)$ (dashed line) in
(2.10) for $\theta_0 = 0.2$, $l_1 = 1$,
$l_2 = 1$, $\Theta_1 = (0.2, 1)$,
$\Theta_2 = (0, 0.2]$

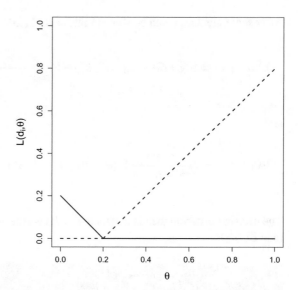

greater (H_1) or not greater (H_2) than a given threshold θ_0. From a decision-theoretic
point of view, two courses of action are possible: d_1 and d_2. Decision d_1 amounts
to accepting the view that the proportion θ is greater than a given (legal) threshold,
θ_0. Decision d_2 amounts to accepting the view that θ is smaller than or equal to the
threshold θ_0. A possible loss function $L(\cdot)$ for such a two-action decision problem
is

$$L(d_1, \theta) = \begin{cases} 0 & \text{if } \theta \in \Theta_1, \\ l_1(\theta_0 - \theta) & \text{if } \theta \in \Theta_2. \end{cases} \qquad L(d_2, \theta) = \begin{cases} 0 & \text{if } \theta \in \Theta_2, \\ l_2(\theta - \theta_0) & \text{if } \theta \in \Theta_1. \end{cases} \qquad (2.10)$$

This is a linear loss function where the loss is proportional to the magnitude of the
error (e.g., $\theta_0 - \theta$). An example is shown in Fig. 2.7, where $\theta_0 = 0.2$, and loss values
l_1 and l_2 are equal to 1.

Given this loss function, the Bayesian posterior expected loss for d_1, that is
accepting $H_1 : \theta > \theta_0$, is

$$\text{EL}(d_1 \mid x) = \int_{\Theta_2} l_1\theta_0 f(\theta \mid x)\mathrm{d}\theta - \int_{\Theta_2} l_1\theta f(\theta \mid x)\mathrm{d}\theta,$$

where $f(\theta \mid x) = \text{Be}(\alpha^* = \alpha + x, \beta^* = \beta + n - x)$. Similarly, the Bayesian
posterior expected loss for d_2, that is accepting $H_2 : \theta \leq \theta_0$, is

$$\text{EL}(d_2 \mid x) = \int_{\Theta_1} l_2\theta f(\theta \mid x)\mathrm{d}\theta - \int_{\Theta_1} l_2\theta_0 f(\theta \mid x)\mathrm{d}\theta.$$

After some algebra, it can be shown (Taroni et al., 2010) that

$$\mathrm{EL}(d_1 \mid x) = l_1 \theta_0 \Pr(\theta < \theta_0 \mid \alpha^*, \beta^*) - l_1 \frac{\alpha + x}{\alpha + \beta + n} \Pr(\theta < \theta_0 \mid \alpha^* + 1, \beta^*),$$

(2.11)

and

$$\mathrm{EL}(d_2 \mid x) = l_2 \frac{\alpha + x}{\alpha + \beta + n} \Pr(\theta > \theta_0 \mid \alpha^* + 1, \beta^*) - l_2 \theta_0 \Pr(\theta > \theta_0 \mid \alpha^*, \beta^*).$$

(2.12)

The decision criterion then is to decide d_1 (d_2) whenever $\mathrm{EL}(d_1)$ is smaller (greater) than $\mathrm{EL}(d_2)$.

Example 2.8 (Counterfeit Medicines—Continued) Recall Example 2.1 where the competing propositions refer to the proportion of counterfeit medicines that may be either greater or not greater than a given limiting value, e.g., $\theta_0 = 0.2$. Consider a uniform prior $\mathrm{Be}(1, 1)$ for θ and the finding that 12 out 40 items are positive. Consider a linear loss function as in (2.10), with $l_1 = 1$ and $l_2 = 1$. This is a symmetric loss, reflecting the idea that falsely deciding that the proportion is greater than the threshold is as undesirable, and hence as severely penalized, as falsely deciding that the proportion is smaller than the threshold. The expected losses of decisions d_1 and d_2 are computed as in (2.11) and (2.12).

```
> th0=0.2
> a=1
> b=1
> n=40
> x=12
> l1=1
> l2=1
> ax=(a+x)/(a+b+n)
> ELd1=l1*th0*pbeta(th0,a+x,b+n-x) -
+ l1*ax*pbeta(th0,a+x+1,b+n-x)
> ELd2=l2*ax*pbeta(th0,a+x+1,b+n-x,lower.tail=F) -
+ l2*th0*pbeta(th0,a+x,b+n-x,lower.tail=F)
> c(ELd1,ELd2)

[1]  0.001207984 0.110731793
```

(continued)

Example 2.8 (continued)
The optimal decision thus is d_1, since it minimizes the expected loss. Given prior beliefs, the observed data, and personal loss assignments, the optimal course of action is to decide in favor of proposition H_1 according to which the proportion of counterfeit medicines is greater than 0.2.

A decision maker may find a "$0 - l_i$" loss function, as shown in Table 1.4, more appropriate. Consider again the case discussed in Sect. 2.2.1 where it was of interest to compare the hypotheses that the proportion of counterfeit medicines in a seizure was greater (H_1) or not greater (H_2) than a given threshold θ_0. In such a context, the loss l_1 (i.e., the loss incurred when deciding d_1 and H_2 is true) could amount to the net loss represented by expenses incurred by issuing legal proceedings in a non-priority case (i.e., falsely considering $\theta > \theta_0$). In turn, loss l_2 could amount to monetary value of property that could have been confiscated by investigative authorities in a meritorious case. Following results in Sect. 1.9, the decision criterion becomes

$$\text{decide } d_1 \text{ if } \quad \frac{\alpha_1}{\alpha_2} > \frac{l_1}{l_2} \quad \text{ or } \quad \text{BF} > \frac{l_1/l_2}{\pi_1/\pi_2}.$$

Decision d_1 is to be preferred to decision d_2 if and only if the posterior odds in favor of H_1 are greater than the ratio of the losses of adverse outcomes or, alternatively, if the BF is greater than the ratio between the loss ratio of adverse outcomes and the prior odds.

Decision makers may find it difficult to assign losses l_1 and l_2. Note, however, that when adverse outcomes are considered equally undesirable, then the loss ratio simplifies to 1, and the decision criterion becomes to decide d_1 whenever the posterior odds are larger than 1, i.e., the posterior probability of hypothesis H_1 is greater than the posterior probability of hypothesis H_2. In turn, when adverse consequences are not equally undesirable, a decision maker may consider how much more (less) undesirable one adverse outcome is compared to the other. This can be expressed as $l_1 = kl_2$, i.e., by specifying how much worse deciding d_1 is when $\theta \leq \theta_0$ is true, compared to deciding d_2 when $\theta > \theta_0$ is true (Biedermann et al., 2016b). A sensitivity analysis can be performed for different values of k.

2.3 Normal Mean

Toxicology laboratories are frequently asked to quantify the amount of target substance (e.g., alcohol, illegal drugs, particular metabolites, etc.) in samples such as blood, urine, and hair in order to help assess whether an unknown target quantity

θ (e.g., the level of alcohol in blood) exceeds a given value (e.g., a legal threshold). Competing propositions of interest may be specified as follows:

H_1:   The target quantity θ exceeds a given level θ_0.
H_2:   The target quantity θ is equal to or smaller than a given level θ_0.

This section considers three main topics: (1) inference about an unknown quantity θ (Sect. 2.3.1), (2) inference about θ in presence of factors influencing the measurement process (Sect. 2.3.2), and (3) decision about competing propositions regarding θ (Sect. 2.3.3).

2.3.1   Inference About a Normal Mean

Consider the hypothetical case of a person, Mr. X, stopped by traffic police because of suspicion of driving under the influence of a given substance (e.g., alcohol or THC). A blood sample is taken and a series of analyses are performed by a forensic laboratory. The propositions of interest may be, for example, that "The quantity θ of target substance in Mr. X's blood exceeds the legal threshold θ_0" (H_1) versus the alternative proposition "The quantity θ of target substance in Mr. X's blood is smaller than or equal to the legal threshold θ_0" (H_2). A series of measurements x are obtained. It is often reasonable to assume that such measurements follow a Normal distribution $N(\theta, \sigma^2)$:

$$f(x \mid \theta, \sigma^2) = \frac{1}{\sqrt{2\pi\sigma^2}} \exp\left\{-\frac{1}{2\sigma^2}(x - \theta)^2\right\},$$

where the mean θ is the unknown quantity of target substance. The variance σ^2 can be approximated from previous ad hoc calibrations (see discussion by Howson and Urbach (1996)). The most common prior distribution for the Normal mean θ is itself a Normal distribution $N(\mu, \tau^2)$:

$$f(\theta \mid \mu, \tau^2) = \frac{1}{\sqrt{2\pi\tau^2}} \exp\left\{-\frac{1}{2\tau^2}(\theta - \mu)^2\right\},$$

where the hyperparameters μ and τ^2 are often called *prior mean* and *prior variance*, respectively.

The posterior distribution of the target quantity θ is still a Normal distribution, denoted $N(\mu_x, \tau_x^2)$, because the Normal prior and the Normal likelihood are conjugate. Generalizing the updating formulae (1.19) and (1.20) to the case where a vector of n measurements $(x_1, \ldots, x_n)$ is available leads to

$$\mu_x = \frac{\sigma^2/n}{\sigma^2/n + \tau^2}\mu + \frac{\tau^2}{\sigma^2/n + \tau^2}\bar{x} \qquad (2.13)$$

and

$$\tau_x^2 = \frac{\tau^2 \sigma^2 / n}{\sigma^2 / n + \tau^2}, \tag{2.14}$$

where $\bar{x} = \sum_{i=1}^n x_i / n$.

The posterior mean μ_x and the posterior variance τ_x^2 can be calculated by means of the function `post_distr`.

```
> post_distr=function(sigma,n,barx,pm,pv){
+          postm=(pm*sigma/n+barx*pv)/(sigma/n+pv)
+          postv=(pv*sigma/n)/(sigma/n+pv)
+          op=c(postm,postv)
+          return(op)}
```

The prior odds, the posterior odds, and the Bayes factor can be easily computed, as discussed in Sect. 1.4, by means of standard routines (see Example 2.9). The case where the population variance σ^2 is unknown and a prior distribution must be specified for both parameters (θ, σ^2) will be addressed in Sect. 3.3.2.

Example 2.9 (Alcohol Concentration in Blood) A person is stopped by traffic police because of suspicion of driving under the influence of alcohol. Two measurements are obtained by the laboratory, 0.4866 g/kg and 0.5078 g/kg. The population variance σ^2 is known and is taken to be equal to 0.023^2. Available information, e.g., the fact that the person has been stopped by traffic police while driving late in the night, exceeding the speed limit etc., suggests a prior mean equal to 0.8 and a prior variance equal to 0.15^2, say $\theta \sim N(\mu = 0.8, \tau^2 = 0.15^2)$. This amounts to say that, a priori, values for the alcohol level in blood lower than 0.35 and larger than 1.25 are considered extremely implausible (prior probabilities for values outside this range are on the order of 0.01).

The propositions of interest are the following:

H_1: The alcohol level θ in the blood of Mr. X exceeds the legal threshold $\theta_0 = 0.5$ ($\theta > 0.5$).

H_2: The alcohol level in the blood of Mr. X is smaller than or equal to the legal threshold $\theta_0 = 0.5$ ($\theta \leq 0.5$).

The prior odds can be easily computed as follows:

(continued)

Example 2.9 (continued)
```
> th0=0.5
> pm=0.8
> pv=0.15^2
> pi1=pnorm(th0,pm,sqrt(pv),lower.tail=F)
> prior_odds=pi1/(1-pi1)
> prior_odds

[1] 42.95579
```

The probability of hypothesis H_1 is, a priori, approximately 43 times greater than the probability of the alternative hypothesis H_2. Consider now the effect of the measurements made on the blood sample.

```
> x=c(0.4866,0.5078)
> s2=0.023^2
> postm=post_distr(s2,length(x),mean(x),pm,pv)[1]
> postm

[1] 0.5007182

> postv=post_distr(s2,length(x),mean(x),pm,pv)[2]
> postv

[1] 0.0002614268
```

The posterior distribution of the quantity of alcohol in blood θ is, therefore, $N(0.5007, 3e - 04)$. The posterior odds are

```
> alpha1=pnorm(th0,postm,sqrt(postv),lower.tail=F)
> post_odds=alpha1/(1-alpha1)
> post_odds

[1] 1.073465
```

The ratio between posterior and prior odds gives the Bayes factor:

```
> BF=post_odds/prior_odds
> BF

[1] 0.02498999
```

The probability to obtain the two measurements if Mr X's alcohol level in blood does *not* exceed the legal threshold $\theta_0 = 0.5$ is approximately 40 times greater than given the proposition that the blood alcohol level is greater than the legal threshold. The evidence thus provides moderate support for the hypothesis H_2, compared to H_1.

2.3.1.1 Choosing the Parameters of the Normal Prior for the Mean

If the experimenter has no reason to consider the distribution describing prior uncertainty about the unknown quantity θ to be asymmetric, then a choice may be made in the family of Normal distributions. When choosing a member from this family, the analyst will need to assign a value to the prior mean μ and a value to the prior standard deviation τ. To elicit a Normal prior, it is useful to recall that for a Normal distribution $\theta \sim N(\mu, \tau^2)$, approximately 99.7% of values are within 3 standard deviation from the mean, thus

$$\Pr\{\mu - 3\tau \leq \theta \leq \mu + 3\tau\} \approx 0.997.$$

Hence, if the practitioner can assign a measure of location μ and a pair of values that define the upper and lower bounds of an interval that covers a range of plausible values of the unknown quantity θ, then the standard deviation can be assigned as

$$\tau = \frac{l_{\text{up}} - \mu}{3}, \tag{2.15}$$

where l_{up} is the upper bound mentioned above. In Example 2.9, a prior location was fixed at $\mu = 0.8$. Moreover, prior probabilities for values smaller than 0.35 and greater than 1.25 were extremely small (i.e., on the order of 0.01). The standard deviation has been elicited as in (2.15).

It may be worth to inspect the reasonableness of the elicited prior. This includes, as highlighted in Sect. 1.10, producing a graphical representation to see whether the amount of available information is suitably conveyed. Consider a random sample of size n_e from a Normal population providing an equivalent amount of information conveyed by the prior. The equivalent sample size n_e can be found by matching the prior variance τ^2 to the dispersion from the sample, σ^2/n_e, and solving for n_e. The smaller n_e, the weaker will be prior beliefs, and the more the posterior distribution will be influenced by even a modest amount of data. Vice versa, the larger n_e, the stronger will be the prior beliefs, and the more the posterior distribution will be dominated by the prior. Thus, more data will be necessary to make a substantial impact on prior beliefs.

Whenever the state of information is such as to consider all possible values of θ equally plausible, a locally uniform prior can be defined:

$$f(\theta) \propto \text{constant}.$$

In the latter case, the posterior distribution of θ is a Normal distribution centered at the sample mean $\bar{x}$ with spread parameter equal to σ^2/n (e.g., Bolstad & Curran, 2017).

2.3.1.2 Sensitivity to the Choice of the Prior Distribution

As noted in Sect. 1.11, the marginal likelihood is highly sensitive to the choice of the prior distribution and so is the Bayes factor. Thus, it should be emphasized that the BF obtained in Example 2.9, the value 0.02, does not depend on the data alone. It also depends on the choice of the prior distribution on θ.

For the purpose of illustration, consider a sensitivity analysis for the hyperparameters that characterize the prior distribution for the unknown level of alcohol in blood. Let values of μ range from 0.4 to 1 and the prior variance τ^2 be fixed and equal to 0.0225.

```
> pm=seq(0.4,1,0.01)
> pv=0.0025
```

The prior odds, the posterior odds, and the BF can be calculated for all possible values of the prior mean μ (pm). Note that computing the posterior Normal distribution with the function post_distr, using several possible values for the prior mean μ, returns an output vector of length $n = 61$ whose first $n - 1 = 60$ elements represent the posterior mean, while the last element represents the posterior variance.

```
> th0=0.5
> pi1=pnorm(th0,pm,sqrt(pv),lower.tail=F)
> prior_odds=pi1/(1-pi1)
> x=c(0.4866,0.5078)
> s2=0.023^2
> postm=
+ post_distr(s2,length(x),mean(x),pm,pv)[1:length(pm)]
> postv=post_distr(s2,length(x),mean(x),pm,pv)
+ [length(pm)+1]
> alpha1=pnorm(th0,postm,sqrt(postv),lower.tail=F)
> post_odds=alpha1/(1-alpha1)
> BF=post_odds/prior_odds
```

Figure 2.8 shows the prior probability π_1 of proposition H_1, the posterior probability α_1, and the BF in favor of proposition H_1 for values of the prior mean μ ranging from 0.4 to 1.

```
> plot(pm,BF,type='l',ylim=c(0,max(pi1,alpha1,BF)),
+ xlim=range(pm),xlab=expression(paste(mu)),ylab='')
```

```
> lines(pm,pi1,lty=4)
> lines(pm,alpha1,lty=2)
> leg=expression(paste('BF'),paste(pi)[1],paste(alpha)
+ [1])
> legend(0.85,1.92,leg,lty=c(1,4,2))
```

Note that the BF favors proposition H_1 (i.e., a BF greater than 1) over H_2 only for values of μ smaller than 0.47. Most importantly, one can observe the impact of the prior assessments (i.e., different choices of the prior mean μ) on the value of the BF. The higher the prior probability of proposition H_1, the lower is the value of the measurements $x = (0.4866, 0.5078)$ in terms of the BF in favor of H_1 over H_2 Note, however, that the BF in the latter case represents strong support for H_2 over H_1.

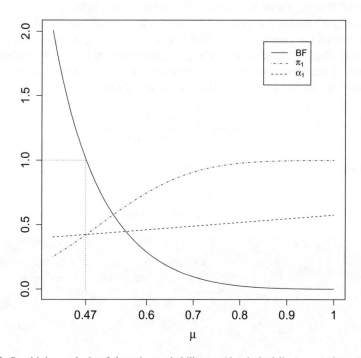

Fig. 2.8 Sensitivity analysis of the prior probability π_1 (dot-dashed line), posterior probability α_1 (dashed line), and BF (solid line) for values of μ ranging from 0.4 to 1 and $\tau^2 = 0.0225$ (Example 2.9). Note that for a BF of 1 (dotted line), the lines of the prior and posterior probabilities intersect

2.3.2   Continuous Measurements Affected by Errors

As noted in Sect. 2.2.2, a measurement process or observations may be affected by background noise. Consider a case in which it is of interest to assess the height of an individual based on video recordings made by a surveillance camera during a bank robbery. Propositions of interest may be as follows:

H_1:   The height of the individual is less than 180 cm.
H_2:   The height of the individual is equal to or greater than 180 cm.

Assume that the height measurements x of an individual are normally distributed, $X \sim N(\theta, \sigma^2)$, where θ represents the true height of the individual and σ^2 represents the variance of the measurement device. Assume also that the variance σ^2 is inferred from previous ad hoc experiments. However, the measured height is, generally, affected by an error ξ, related to the circumstances under which the recording was made. Factors of interest here include the posture and movements of the person, the type of clothing (including headwear and shoes) and lighting conditions. Such circumstances represent a further source of variation δ^2, unrelated to σ^2. The measured height is therefore $X \sim N(\theta + \xi, \sigma^2 + \delta^2)$. A conjugate Normal prior distribution $N(\mu, \tau^2)$ is taken to model prior uncertainty about θ. The values of the parameters ξ and δ^2 are case-specific assignments. It can be shown that the posterior distribution of the true height θ is still Normal with mean

$$\mu_x = \frac{\tau^2(\bar{x} - \xi) + \mu(\sigma^2 + \delta^2)/n}{\tau^2 + (\sigma^2 + \delta^2)/n} \tag{2.16}$$

and variance

$$\tau_x^2 = \frac{\tau^2(\sigma^2 + \delta^2)/n}{\tau^2 + (\sigma^2 + \delta^2)/n}. \tag{2.17}$$

Example 2.10 (Image Analysis)   Consider the hypothetical case introduced above and assume that, according to eyewitness testimony, the height of the perpetrator is approximately between 175 cm and 185 cm. This allows one to define a prior probability distribution for the height θ centered at 180 cm with variance equal to 2.79 cm, i.e., $\theta \sim N(180, 2.79)$. The standard deviation can be quantified as in (2.15):

```
> lsup=185
> pm=180
> ps=(lsup-pm)/3
> pv=ps^2
```

(continued)

Example 2.10 (continued)

Thus, the two hypotheses H_1 and H_2 introduced above are, a priori, equally probable (hence, the prior odds equal 1).

```
> th0=180
> pi1=pnorm(th0,pm,sqrt(pv))
> prior_odds=pi1/(1-pi1)
> prior_odds

[1] 1
```

The available recordings depict an individual appearing in $n = 10$ images. Height measurements yield the sample mean $\bar{x} = 180.25$. The variance of the measurement procedure is known and equal to $\sigma^2 = 0.12$. The experimental setting is such that the values for the parameters of the Normal distribution of the error can be set to $\xi = 0.5$ and $\delta^2 = 1$.

```
> mx=180.25
> n=10
> s2=0.12
> xi=0.5
> d2=1
```

The posterior mean and the posterior variance of θ can be computed as in (2.16) and (2.17), respectively.

```
> postm=(pv*(mx-xi)+pm*(s2+d2)/n)/(pv+(s2+d2)/n)
> postm

[1] 179.7597

> postv=(pv*(s2+d2)/n)/(pv+(s2+d2)/n)
> postv

[1] 0.1076592
```

The gray shaded area in Fig. 2.9 shows the posterior probability of the hypothesis H_1. The posterior odds and the Bayes factor can be obtained straightforwardly

```
> alpha1=pnorm(th0,postm,sqrt(postv))
> post_odds=alpha1/(1-alpha1)
> post_odds

[1] 3.311039

> BF=post_odds/prior_odds
> BF

[1] 3.311039
```

(continued)

Fig. 2.9 Posterior
distribution $f(\theta \mid x)$ for the
true height θ in
Example 2.10. The gray
shaded area shows the
posterior probability of the
hypothesis H_1 ($\theta < 180$ cm)

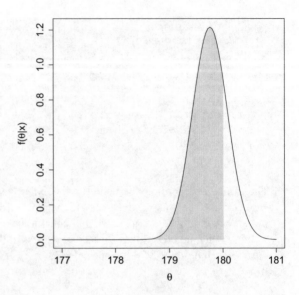

Example 2.10 (continued)

Given that the prior odds are 1, the BF is numerically equivalent to the posterior odds. This value represents support for the hypothesis H_1 (the height of the individual is lower than 180 cm) over H_2. Specifically, the BF indicates that it is approximately 3 times more probable to obtain such height measurements if the height of the individual is less than 180 cm than if the height is equal to or greater than 180 cm.

2.3.3   Decision for a Mean

The previous sections focused on how to draw a probabilistic inference about a Normal mean, using the Bayes factor. Recall that the competing propositions were:

H_1:   The target quantity θ exceeds a given level θ_0.
H_2:   The target quantity θ is equal to or smaller than a given level θ_0.

A related question is how to *decide* about whether or not a quantity of interest is above a given (legal) threshold, i.e., accepting either H_1 or H_2. In order to address this question, it is necessary to introduce a loss function to take into account the decision maker's preferences. Suppose a linear loss function is considered as in (2.18):

$$L(d_1, \theta) = \begin{cases} 0 & \text{if } \theta > \theta_0, \\ l_1(\theta_0 - \theta) & \text{if } \theta \le \theta_0. \end{cases} \qquad L(d_2, \theta) = \begin{cases} 0 & \text{if } \theta \le \theta_0, \\ l_2(\theta - \theta_0) & \text{if } \theta > \theta_0. \end{cases} \quad (2.18)$$

The Bayesian posterior expected loss of decision d_1 can be computed as

$$\text{EL}(d_1 \mid x) = l_1 \int_{\theta \le \theta_0} (\theta_0 - \theta) f(\theta \mid x) d\theta$$

$$= l_1 \tau_x \left[\phi(t) + t \int_0^t \phi(s) ds \right], \quad (2.19)$$

where $f(\theta \mid x)$ is a Normal posterior distribution with parameters μ_x and $\tau^2(x)$ as in (2.13) and (2.14), $t = \tau_x(\theta_0 - \mu_x)$, while $\phi(\cdot)$ denotes the probability density of a standardized Normal distribution (Bernardo & Smith, 2000).

In turn, the Bayesian posterior expected loss of decision d_2 can be computed as

$$\text{EL}(d_2 \mid x) = l_2 \int_{\theta > \theta_0} (\theta - \theta_0) f(\theta \mid x) d\theta$$

$$= l_2 \tau_x \left[\phi(t) - t \int_t^\infty \phi(s) ds \right]. \quad (2.20)$$

Again, the decision criterion amounts to deciding d_1 (d_2) whenever $\text{EL}(d_1 \mid x)$ is smaller (greater) than $\text{EL}(d_2 \mid x)$.

Example 2.11 (Alcohol Concentration in Blood—Continued)   Recall Example 2.9 where the posterior distribution of the alcohol level θ was N(0.50072, 0.00026), and the legal threshold was equal to 0.5.

```
> th0=0.5
> postm

[1]  0.5007182

> postv

[1]  0.0002614268
```

Consider a symmetric linear loss function as in (2.18) with $l_1 = l_2 = 1$. The Bayesian posterior expected losses in (2.19) and (2.20) can be obtained as

```
> l1=1
> l2=1
> t=sqrt(postv)*(th0-postm)
```

(continued)

Example 2.11 (continued)
```
> eld1=l1*sqrt(postv)*(dnorm(t)+t*(pnorm(t)-0.5))
> eld2=l2*sqrt(postv)*(dnorm(t)-t*pnorm(t,lower.
+ tail=F))
> c(eld1,eld2)

[1] 0.006450377 0.006450471
```

The optimal decision thus is to consider that the alcohol level is greater than the legal threshold because this decision has a lower expected loss, though the difference between the two expected losses is, in the example here, extremely small

```
> abs(eld1-eld2)

[1] 9.388144e-08
```

Note that this result crucially depends on the decision maker's value assessments (i.e., the chosen loss function).

When expected losses for rival decisions are very similar, as is the case in Example 2.11, a sensitivity analysis should be performed as suggested, for example, in legal literature (Edwards, 1988). The sensitivity analysis should evaluate the effect of changes in the prior parameters and the loss values. See also Sect. 2.3.1 for a sensitivity analysis of the BF for evaluating the impact of changes in hyperparameters characterizing the prior distribution for the unknown level of alcohol in blood.

It is also worth to reflect on the choice of the loss function. A symmetric loss function, as previously suggested, may not realistically reflect the decision maker's preferences. For example, a decision maker who is concerned about road safety may consider that falsely concluding that an individual's blood alcohol concentration is below the legal limit is a more serious error than falsely concluding that an individual's blood alcohol concentration is above the legal threshold. Therefore, l_2 may be taken to be larger than l_1, reflecting the greater inconvenience associated with underestimating the alcohol concentration. For example, when $l_1 = 1$ and $l_2 = 2$, meaning that underestimating the alcohol level is considered twice as serious as overestimating it, the expected loss of decision d_2 will increase. One can verify that for any reasonable value of l_2 greater than l_1, decision d_1 will be the one with the smaller expected loss.

2.4 Summary of R Functions

The R functions outlined below have been used in this chapter.

Functions Available in the Base Package

apply: applies a function to the margins (either rows or columns) of a matrix

acf: computes and plots estimates of the autocorrelation function

d<name of distribution>, p<name of distribution>,
r<name of distribution> (e.g., dbeta, pbeta, rbeta): calculates the
  density and the cumulative probability and generates random numbers for various
  parametric distributions

rowSums: forms row sums for numeric arrays (or data frames)

Further details can be found in the Help menu, help.start().

Functions Available in Other Packages

dbbinom and pbbinom in package extraDistr: calculates the density and the
cumulative probability for a beta-binomial distribution

Functions Developed in This Chapter

dbinpois: computes the product between a binomial likelihood $\mathrm{Bin}(n, \theta)$ at $x -$
x_b and a Poisson likelihood $\mathrm{Pn}(\lambda)$ at x_b where x represents the number of items
counted as presenting a given target characteristic and x_b represents the number of
background elements affecting the counting process

Usage: dbinpois(xb)
Arguments: xb: a vector of integers ranging from 0 to x
Output: a vector of values, where each value represents the probability of the product
  between the binomial and the Poisson likelihood at a given value of the input
  argument xb

mh1: computes the function $x/(1 + x)$
Usage: mh1(x)
Arguments: x: a scalar value x
Output: the value of $x/(1 + x)$

mh2: computes the function $x/(1 + x)^2$
Usage: mh2(x)
Arguments: x: a scalar value x
Output: the value of $x/(1 + x)^2$

post_distr: computes the posterior distribution $N(\mu_x, \tau_x^2)$ of a Normal mean θ, with $X \sim N(\theta, \sigma^2)$ and $\theta \sim N(\mu, \tau^2)$

Usage: post_distr(sigma,n,barx,pm,pv)

Arguments: sigma, the variance σ^2 of the observations; n, the number of observations; barx, the sample mean $\bar{x}$ of the observations; pm, the mean μ of the prior distribution $N(\mu, \tau^2)$ and pv, the variance τ^2 of the prior distribution $N(\mu, \tau^2)$

Output: a vector of two values: the first is the posterior mean μ_x and the second is the posterior variance τ_x^2

Published with the support of the Swiss National Science Foundation (Grant no. 10BP12_208532/1).

Chapter 3
Bayes Factor for Evaluative Purposes

3.1 Introduction

Consider a case where material of known source (control material) and evidential material of unknown source (recovered or questioned material) are collected and analyzed. Interpretation of scientific evidence then amounts to assessing the probative value of the observations made during comparative examinations. The evidence is evaluated in terms of its effect on the odds in favor of a proposition H_1 put forward by the prosecution, compared to an alternative proposition H_2 advanced by the defense.

During comparative examinations, observations and measurements are made, leading to either discrete or continuous data. Forensic laboratories may also have equipment and methodologies that can lead to output in the form of multivariate data. Thus, scientific evidence is often described by more than one variable. For example, glass fragments from a crime scene can be compared with fragments collected on the clothing of a person of interest on the basis of several chemical components, as well as physical characteristics. It should be noted, however, that the assessment of a Bayes factor for multivariate data may be challenging. For example, data may not present enough regularity so that standard parametric distributions cannot be used. Data may also present a complex dependence structure with several levels of variation. In addition, a feature-based approach might not be always feasible, and it may be necessary to derive a Bayes factor on the basis of scores.

This chapter is structured as follows. Sections 3.2 and 3.3 address the problem of evaluation of evidence for various types of discrete and continuous data, respectively. Section 3.4 presents an extension to continuous multivariate data.

Supplementary Information The online version contains supplementary material available at https://doi.org/10.1007/978-3-031-09839-0_3. The files can be accessed individually by clicking the DOI link in the accompanying figure caption or by scanning this link with the SN More Media App.

© The Author(s) 2022
S. Bozza et al., *Bayes Factors for Forensic Decision Analyses with R*,
Springer Texts in Statistics, https://doi.org/10.1007/978-3-031-09839-0_3

3.2   Evidence Evaluation for Discrete Data

This section deals with measurement results in the form of counts, using the binomial model (Sect. 3.2.1), the multinomial model (Sect. 3.2.2), and the Poisson model (Sect. 3.2.3).

3.2.1   Binomial Model

In many practical applications, data derive from realizations of experiments that may take one of two mutually exclusive outcomes. Examples include general features (so-called class characteristics) observed on questioned and known items or materials (e.g., fired bullets, fibers) when the question of interest is whether the compared materials come from the same source.

Consider a hypothetical case involving a questioned document for which results of analyses of black toner are available. On the questioned document, black bi-component toner is present. It is of the same type as that used by a given printing machine (known source). A question that may be of interest in such a case is how this analytical information should affect one's belief in the proposition according to which the questioned document has been printed using the device of interest (Biedermann et al., 2009, 2011a). The competing propositions can thus be defined as follows:

H_1 :   The questioned document has been printed with the device of interest.
H_2 :   The questioned document has been printed with an unknown device.

Let T denote the observed toner type, either single component (T_S) or bi-component (T_B). Suppose that a database of the toner type (magnetism) of samples of black toner from N machines is available, n of which use a bi-component toner. Denote by θ the proportion of the population of printing devices equipped with bi-component toner. Available counts can be treated as realizations of Bernoulli trials (Sect. 2.2.1) with constant probability of success θ, $\Pr(T_B \mid \theta) = \theta$. Suppose a conjugate beta prior distribution $\mathrm{Be}(\alpha, \beta)$ is used to model uncertainty about θ, where α and β can be elicited using the available background knowledge as in (1.42) and (1.43).

Denote by E_y the observations made on recovered material and by E_x the observations made on control material (i.e., documents printed with the device of interest). If the questioned document originates from the device of interest, the probability of the evidence becomes

$$\Pr(E_y = T_B, E_x = T_B \mid H_1) = \int_\Theta \Pr(T_B \mid \theta) \cdot \theta^{\alpha-1}(1-\theta)^{\beta-1} d\theta / \mathrm{B}(\alpha, \beta)$$

$$= \int_\Theta \theta \cdot \theta^{\alpha-1}(1-\theta)^{\beta-1} d\theta / \mathrm{B}(\alpha, \beta).$$

If the questioned document originates from an unknown device (i.e., two distinct devices have been used), the probability of the evidence becomes

$$\Pr(E_y = T_B, E_x = T_B \mid H_2) = \int_\Theta \theta^2 \cdot \theta^{\alpha-1}(1-\theta)^{\beta-1}d\theta/\mathrm{B}(\alpha, \beta).$$

The Bayes factor can be computed as

$$\begin{aligned}
\mathrm{BF} &= \frac{\int_\Theta \theta \cdot \theta^{\alpha-1}(1-\theta)^{\beta-1}d\theta}{\int_\Theta \theta^2 \cdot \theta^{\alpha-1}(1-\theta)^{\beta-1}d\theta} \\
&= \frac{\mathrm{B}(\alpha+1, \beta)}{\mathrm{B}(\alpha+2, \beta)} \int_\Theta \frac{\theta^\alpha(1-\theta)^{\beta-1}}{\theta^{\alpha+1}(1-\theta)^{\beta-1}} \frac{\mathrm{B}(\alpha+2, \beta)}{\mathrm{B}(\alpha+1, \beta)} \\
&= \frac{\alpha+\beta+1}{\alpha+1}.
\end{aligned} \tag{3.1}$$

Example 3.1 (Questioned Documents) Consider the case of a printed document of unknown origin. Analyses reveal that the toner present on the printed document is of type "bi-component." The printing device that is thought to have been used to print the questioned document is equipped with a bi-component toner. In an available database with a total of $N = 100$ samples of black toner, $n = 23$ are bi-component (see Table 3.1). Using this information, the parameters of the beta prior distribution about θ can be elicited as follows:

```
> n=23
> N=100
> p=n/N
> a=p*(N-1)
> b=(1-p)*(N-1)
```

This leads to a Be(23, 76).

The Bayes factor in (3.1) can be computed straightforwardly as follows:

```
> BF=(a+b+1)/(a+1)
> BF
```

```
[1] 4.206984
```

The Bayes factor provides weak support for the proposition H_1 according to which the questioned document has been printed with the printing device of interest rather than with an unknown printing device (H_2).

It is worth noting that there is an alternative development described in the forensic statistics literature that considers background information derived from a population database as part of the evidence, (e.g., Ommen et al., 2016; Dawid,

Table 3.1 Results obtained following the analysis of, respectively, the component type (magnetism) and the resin type of 100 samples of black toner (Biedermann et al., 2011a)

Resin group	Single component	Bi-component
1. Styrene-co-acrylate	69	14
2. Epoxy A	8	3
3. Epoxy B	0	2
4. Epoxy C	0	1
5. Epoxy D	0	1
6. Polystyrene	0	1
7. Other	0	1

2017). According to this line of reasoning, if proposition H_1 is true (numerator), there are $(n + 1)$ counts of bi-component toners. That is, the questioned item and the known item are assumed to come from the same source, hence adding one count to the database. Conversely, if proposition H_2 is true (denominator), there are $(n+2)$ counts of bi-component toner. Here, it is assumed that the questioned item and the known item come from different sources, hence adding two counts to the database. The Bayes factor can then be obtained as

$$
\text{BF} = \frac{\int_\Theta \theta^{n+1}(1-\theta)^{N-n}\theta^{\alpha-1}(1-\theta)^{\beta-1}d\theta}{\int_\Theta \theta^{n+2}(1-\theta)^{N-n}\theta^{\alpha-1}(1-\theta)^{\beta-1}d\theta}
$$
$$
= \frac{\alpha + \beta + N + 1}{\alpha + n + 1}. \tag{3.2}
$$

One can immediately verify that this corresponds to the BF in (3.1) with parameter α replaced by $\alpha + n$, and parameter β replaced by $\beta + N - n$. However, it may be questioned whether the available database should be considered as evidence, rather than as conditioning information, because the database contains only general data unrelated to the case under investigation (Aitken et al., 2021).

3.2.2 Multinomial Model

The analyses described in Sect. 3.2.1 can be extended to situations where experiments can lead to more than two mutually exclusive outcomes.

Consider again the case involving printed documents, introduced in Sect. 3.2.1. Laboratories often analyze resins of toner on printed documents by means of Fourier Infrared Spectroscopy (FTIR). The results can be classified into one of several (k) categories (Table 3.1). Suppose that the resin type (R) recovered on the questioned document belongs to category j, which is also found in the toner used by a given printing machine. The question of interest is similar to the one considered in Sect. 3.2.1, that is, how the available analytical information should affect one's belief in the proposition according to which a questioned document has been printed using a given device, called the potential source, rather than by some unknown printing device.

Denote by θ_j the proportion of the population that is of type (category) R_j, $j = 1, \ldots, k$, $\Pr(R_j \mid \theta_j) = \theta_j$. Assume that observations of distinct categories can be treated as independent: available counts $n_1, \ldots, n_k$ can be treated as realizations from a multinomial distribution $\mathrm{Mult}(n, \theta_1, \ldots, \theta_k)$

$$f(n_1, \ldots, n_k \mid \theta_1, \ldots, \theta_k) = \frac{N!}{n_1! \cdots \cdots n_k!} \theta_1^{n_1} \cdots \cdots \theta_k^{n_k}.$$

A conjugate Dirichlet prior probability distribution $\mathrm{Dir}(\alpha_1, \ldots, \alpha_k)$ is considered for modeling uncertainty about the population proportions $\theta_1, \ldots, \theta_k$:

$$f(\theta_1, \ldots, \theta_k \mid \alpha_1, \ldots, \alpha_k) = \theta_1^{\alpha_1 - 1} \cdots \cdots \theta_k^{\alpha_k - 1} / \mathrm{B}(\alpha),$$

with $\mathrm{B}(\alpha) = \frac{\prod_{i=1}^{k} \Gamma(\alpha_i)}{\Gamma(\alpha)}$ and $\alpha = \sum_{i=1}^{k} \alpha_i$.

Denote by E_y the observations made on the recovered material and by E_x the observations made on the control material (i.e., documents printed with the device of interest). If the questioned document originates from the device of interest, the probability of the findings $E = (E_y, E_x)$ becomes

$$\Pr(E_y = R_j, E_x = R_j \mid H_1) = \int_{\Theta} \Pr(R_j \mid \theta_j) \cdot \theta_1^{\alpha_1 - 1} \cdots \cdots \theta_j^{\alpha_j - 1} \cdots \cdots \theta_k^{\alpha_k - 1} d\theta / \mathrm{B}(\alpha)$$

$$= \int_{\Theta} \theta_j \cdot \theta_1^{\alpha_1 - 1} \cdots \cdots \theta_j^{\alpha_j - 1} \cdots \cdots \theta_k^{\alpha_k - 1} d\theta / \mathrm{B}(\alpha).$$

If the questioned documents originate from an unknown device (i.e., two distinct devices have been used), the probability of the findings E becomes

$$\Pr(E_y = R_j, E_x = R_j \mid H_2) = \int_{\Theta} \theta_j^2 \cdot \theta_1^{\alpha_1 - 1} \cdots \cdots \theta_j^{\alpha_j - 1} \cdots \cdots \theta_k^{\alpha_k - 1} d\theta / \mathrm{B}(\alpha).$$

The Bayes factor can be computed as

$$\mathrm{BF} = \frac{\int \theta_j \cdot \theta_1^{\alpha_1 - 1} \cdots \cdots \theta_j^{\alpha_j - 1} \cdots \cdots \theta_k^{\alpha_k - 1} d\theta}{\int \theta_j^2 \cdot \theta_1^{\alpha_1 - 1} \cdots \cdots \theta_j^{\alpha_j - 1} \cdots \cdots \theta_k^{\alpha_k - 1} d\theta}$$

$$= \frac{\alpha + 1}{\alpha_j + 1}. \tag{3.3}$$

Example 3.2 (Questioned Documents—Continued) Recall Example 3.1, involving questioned documents on which black toner is present. Suppose now that laboratory analyses focus on the toner's resin component. Suppose that the parameters of the Dirichlet prior probability distribution are elicited as

```
> a=c(15,4,3,2,2,2,2)
```

Suppose that the rather common resin group *Epoxy-A* (category $j = 2$ in Table 3.1) is observed on both the questioned and known documents. The Bayes factor in (3.3) can be computed straightforwardly as

```
> j=2
> BF=(sum(a)+1)/(a[j]+1)
> BF

[1] 6.2
```

The Bayes factor provides, again, weak support for the proposition H_1 according to which the questioned document has been printed with the printing device of interest, rather than with an unknown printing device (H_2).

Suppose that a database of the resin type of samples of black toner from N machines is available, n_1 (n_2, ...) of which belong to category 1 (2, ...), as in Table 3.1. These data can be used to elicit the Dirichlet prior probability distribution. Following the methodology proposed by Zapata-Vazquez et al. (2014), the hyperparameters $\alpha_1, \ldots, \alpha_k$ can be assessed by starting from expert judgments (e.g., a vector of quantiles) about proportions of items belonging to each category. Tools for eliciting prior probability distributions from experts' opinions are also available in the R package SHELF. An example will be presented in Sect. 4.2.2.

3.2.3 Poisson Model

Some forensic science applications focus on the number of occurrences of particular events or observations that take place at given intervals of time or space. Practical examples are the number of gunshot residue particles (GSR) collected on the surface of the hands of individuals suspected to be involved in the discharge of a firearm (Cardinetti et al., 2006), or the number of corresponding matching striations in the comparative examination of marks left by firearms on fired bullets (Bunch, 2000).

Consider the following hypothetical case. A fired bullet is found at a crime scene, and a person of interest is apprehended, carrying a gun. The following propositions are of interest:

H_1 :   The bullet found at the crime scene was fired with the seized gun.

H_2 :   The bullet found at the crime scene was fired with an unknown gun.

The recovered bullet and bullets fired with the seized gun are compared. *Consecutive matching striations* (CMS) is a simple concept to quantify the extent of agreement between marks. The number of observed consecutively matching striations can be interpreted as a *score*. Let $\Delta(x, y)$ be the maximum CMS count for a given comparison. For the evaluation of a CMS count, data on comparisons made between pairs of bullets test-fired with the seized gun and between pairs of bullets test-fired with different guns are needed. The (score-based) Bayes factor therefore is

$$\text{sBF} = \frac{g(\Delta(x, y) \mid H_1)}{g(\Delta(x, y) \mid H_2)}.$$

A statistical model commonly used in the forensic science literature for the type of data encountered in the example here assumes that counts follow a Poisson distribution $\text{Pn}(\lambda)$

$$g(\Delta(x, y) \mid \lambda_i) = \frac{e^{-\lambda_i} \lambda_i^{\Delta(x,y)}}{\Delta(x, y)!}, \qquad \Delta(x, y) = 0, 1, \ldots \; ; \; \lambda_i \geq 0,$$

where parameter λ_i, $i = 1, 2$, represents the weighted average maximum CMS count.

Suppose that two datasets are compiled. The first relates to pairs of bullets fired with the seized gun, and the second to pairs of bullets fired with different guns. Such data can be used to inform the probability distribution $g(\cdot)$ at the score value $\Delta(x, y)$ as discussed in Sect. 1.5.2 and to compute the Bayes factor as

$$\text{sBF} = \frac{\hat{g}(\Delta(x, y) \mid x, H_1)}{\hat{g}(\Delta(x, y) \mid H_2)}.$$

Bunch (2000) describes a likelihood ratio procedure for inference about competing propositions. This account is based on a frequentist perspective because it uses the maximum likelihood estimates $\hat{\lambda}_1$ and $\hat{\lambda}_2$ for parameters λ_1 and λ_2, calculated under the assumption that either proposition H_1 or proposition H_2 is true. Using these two estimates in the component Poisson likelihoods leads to the following likelihood ratio:

$$\text{LR} = \frac{e^{-\hat{\lambda}_1} \hat{\lambda}_1^{\Delta(x,y)}}{e^{-\hat{\lambda}_2} \hat{\lambda}_2^{\Delta(x,y)}}.$$

In Bayesian statistics, the most common prior distribution for λ_i is the gamma distribution $\text{Ga}(\alpha_i, \beta_i)$ with shape parameter α and rate parameter β (e.g. Bernardo and Smith, 2000):

$$f(\lambda_i \mid \alpha_i, \beta_i) = \frac{\beta_i^{\alpha_i}}{\Gamma(\alpha_i)} \lambda_i^{\alpha_i - 1} e^{-\beta_i \lambda_i}, \qquad \lambda_i > 0 \; ; \; \alpha_i, \beta_i > 0.$$

Since the Poisson and gamma distributions are conjugate (Sect. 1.10), the posterior distribution of λ is still in the family of gamma distributions, with parameters α and β updated according to well-known updating rules (see, e.g., Lee, 2012). When we have a realization of a random sample from a Poisson distribution, $Pn(\lambda)$, say $(z_1, \ldots, z_n)$, we end up with a $Ga(\alpha', \beta')$, where $\alpha' = \alpha + \sum_{i=1}^{n} z_i$ and $\beta' = \beta + n$. Note that in this case there is only one observation, $\Delta(x, y)$; therefore, $\alpha' = \alpha + \Delta(x, y)$ and $\beta' = \beta + 1$. See also Biedermann et al. (2011b) for further illustrations of the Poisson–gamma model in forensic science applications.

The marginal distribution in the numerator and denominator of the Bayes factor is known in closed form here. It is a Poisson–gamma distribution:

$$g(\Delta(x, y) \mid \alpha_i, \beta_i) = \int_{\lambda_i} g(\Delta(x, y) \mid \lambda_i) f(\lambda_i \mid \alpha_i, \beta_i) d\lambda_i$$

$$= \frac{1}{\Delta(x, y)!} \frac{\beta_i^{\alpha_i}}{\Gamma(\alpha_i)} \frac{\Gamma(\alpha_i + \Delta(x, y))}{(\beta_i + 1)^{\alpha_i + \Delta(x, y)}}. \tag{3.4}$$

The score-based Bayes factor then becomes

$$\text{sBF} = \frac{\beta_1^{\alpha_1} \Gamma(\alpha_2) \Gamma(\alpha_1 + \Delta(x, y)) (\beta_2 + 1)^{\alpha_2 + \Delta(x, y)}}{\beta_2^{\alpha_2} \Gamma(\alpha_1) \Gamma(\alpha_2 + \Delta(x, y)) (\beta_1 + 1)^{\alpha_1 + \Delta(x, y)}}. \tag{3.5}$$

Another example of the use of the Poisson distribution for data in the form of independent counts can be found in Aitken and Gold (2013). These authors considered the number of occurrences of selected characteristics of speech recorded in a succession of time periods. In this application, a feature-based Bayes factor is used to assess findings with respect to the proposition according to which recorded and control speeches originate from the same source versus the alternative proposition that they originate from different sources.

Example 3.3 (Firearm Examination)   Consider a case involving a questioned bullet. During comparison with a reference bullet, the examiner counts four CMS, i.e., $\Delta(x, y) = 4$. Suppose that the assumptions made in Bunch (2000) are suitable for the case here so that for bullets fired from the same gun (proposition H_1 holds), the weighted average maximum CMS is taken to be equal to 3.91. For bullets fired from different guns (proposition H_2 holds), the weighted average maximum CMS count is taken to be equal to 1.32. These values are used in the Poisson likelihoods under H_1 and H_2, and the likelihood ratio can easily obtained as

```
> s=4
> lambda1=3.91
```

(continued)

Example 3.3 (continued)
```
> lambda2=1.32
> LR=dpois(s,lambda1)/dpois(s,lambda2)
> LR
```
[1] 5.775487

The evidence provides weak support in favor of the proposition according to which the recovered bullet passed through the barrel of the seized gun, rather than through the barrel of an unknown gun.

Consider now the Bayesian perspective. Suppose that the available knowledge allows one to set the hyperparameters of the gamma distribution equal to $\{\alpha_1 = 125, \beta_1 = 32\}$ for the numerator and to $\{\alpha_2 = 7, \beta_2 = 5\}$ for the denominator. This amounts to using a gamma prior distribution for λ_1 with mean equal to 3.91 and standard deviation equal to 0.35 and a gamma prior distribution for λ_2 with mean equal to 1.4 and standard deviation equal to 0.53. The two prior distributions are shown in Fig. 3.1.

```
> an=125
> bn=32
> ad=7
> bd=5
> plot(function(x) dgamma(x,an,bn),0,8,
+ xlab=expression(paste(lambda)),ylab='Probability
+ density')
> plot(function(x) dgamma(x,ad,bd),0,8,add=TRUE,
+ lty=2)
> leg=expression(paste('Ga(125,32)'),paste(
+'Ga(7,5)'))
> legend(4.85,1.15,leg,lty=c(1,2))
```

First, we write a short function `poisg` that computes the marginal distribution in (3.4)

```
> poisg=function(a,b,x)
+ {(b^a)/gamma(a)*gamma(a+x)/((b+1)^(a+x))}
```

Next, the Bayes factor can be computed as follows:

```
> BF=poisg(an,bn,s)/poisg(ad,bd,s)
> BF
```

[1] 4.248019

Note that the introduction of a prior probability distribution reflecting uncertainty about the population parameters λ_1 and λ_2 has slightly lowered the value of the evidence. The result still represents weak evidence in favor of the

(continued)

Fig. 3.1 Gamma prior for the
Poisson parameter λ under
H_1 (solid line) and H_2
(dashed line)

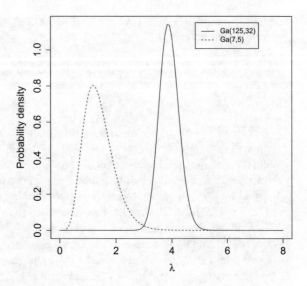

Example 3.3 (continued)
proposition that the recovered bullet was fired with the seized gun, rather than
with an unknown gun.

Note that Example 3.3 involves a non-anchored approach at the numerator. The
probability distribution of the score value is solely conditioned on the hypothesis of
interest, that is $\hat{g}(\Delta(x, y) \mid H_1)$. As mentioned at the beginning of this section, and
in Sect. 1.5.2, other anchoring approaches may be considered.

3.2.3.1   Choosing the Parameters of the Gamma Prior

An evaluator who, initially, would like to give the same weight to all possible values
of λ may consider to use a non-informative prior distribution, that is

$$f(\lambda_i) = \lambda_i^{-1/2}; \qquad \lambda_i > 0 \text{ and } i = 1, 2.$$

The posterior probability distribution given the observations $(z_1, \ldots, z_n)$ will be
of type gamma with shape parameter $\alpha' = \sum_{i=1}^{n} z_i + 1/2$ and rate parameter
$\beta' = n$. Note that in the type of case considered here, there is only one observation;
therefore, $\alpha' = \Delta(x, y) + 1/2$ and $\beta' = 1$.

However, the choice of a non-informative prior distribution may be questioned.
Take, for instance, the case example discussed earlier in this section (Example 3.3).
It is difficult to imagine that *no* suitable information is available to express prior

uncertainty about the unknown weighted average maximum count CMS, and hence that the same non-informative prior distribution should apply under each proposition.

In Example 3.3, an informative prior distribution has been used. This raises the question of how to translate prior knowledge into a prior distribution. As illustrated in Sect. 1.10, one way to elicit prior parameters is to express prior beliefs in terms of a measure of location and a measure of dispersion and then equate these values with the prior moments of the distribution. In the case of a gamma distribution $Ga(\alpha, \beta)$, this amounts to equate a value for the mean, m, with the prior mean α/β, and a value for the variance, s^2, with the prior variance α/β^2, that is,

$$m = \frac{\alpha}{\beta} \quad ; \quad s^2 = \frac{\alpha}{\beta^2}.$$

Solving for α and β gives

$$\alpha = \frac{m^2}{s^2} \tag{3.6}$$

$$\beta = \frac{m}{s^2}. \tag{3.7}$$

If the shape of the prior distribution resulting from the choice of α and β as in (3.6) and (3.7) does not reflect one's prior beliefs suitably, then one should adjust the numerical values of m and s. However, this may not be enough to ensure that the resulting prior distribution is reasonable. One should also inquire about whether the information that is conveyed by the prior is realistically attainable. Consider a random sample of size n_e, providing the same amount of information as conveyed by the elicited prior. The sample mean should have, at least roughly, the same location and the same dispersion as the prior. The equivalent sample size n_e can then be found by matching the moments of the gamma distribution to the corresponding moments characterizing a sample of size n_e from a Poisson distributed random variable located at λ:

$$\frac{\alpha}{\beta} = \lambda$$

$$\frac{\alpha}{\beta^2} = \frac{\lambda}{n_e}.$$

If the mean λ is set equal to the prior mean α/β, the equivalent sample size n_e is equal to β.

Example 3.4 (Elicitation of a Gamma Prior) In Example 3.3, a Ga(125, 32) was used for λ_1 (the weighted average maximum CMS count under proposition H_1), and a Ga(7, 5) for λ_2 (the weighted average maximum CMS count under proposition H_2). For the prior means of λ_1 and λ_2, the values 3.91 and 1.4 were used following Bunch (2000). For the dispersion of the two distributions, the values 0.35 and 0.53 have been assigned to the standard deviation under propositions H_1 and H_2, respectively. Parameters ($\alpha_1 = 125, \beta_1 = 32$) and ($\alpha_2 = 7, \beta_2 = 5$) have then been obtained as in (3.6) and (3.7). This amounts to an equivalent sample size equal to 32 for the prior density of λ_1, and 5 for λ_2.

3.2.3.2 Sensitivity to Prior Probabilities of Competing Propositions

It is important to emphasize that the analyses presented here make no direct probabilistic statement about the truth of the propositions put forward by opposing parties at trial. A Bayes factor of approximately 4.25, as obtained in Example 3.3, only means that the evidence is approximately 4 times more probable if proposition H_1 is true than if the alternative proposition H_2 is true. As noted earlier, this does not mean that proposition H_1 is more probable than H_2. This depends on the prior probabilities of the competing propositions, which can vary considerably among recipients of expert information, and which are beyond the area of competence of scientists.

However, it may be of interest to show the impact of different prior probability assignments on the posterior probability of the competing propositions. To do so, recall that the posterior odds are given by the product of the prior odds and the Bayes factor

$$\frac{\Pr(H_1 \mid \cdot)}{\Pr(H_2 \mid \cdot)} = \text{BF} \times \frac{\Pr(H_1)}{\Pr(H_2)}.$$

Using this expression, one can then investigate how the posterior probability of proposition H_1, i.e., α_1, varies for values of π_1, i.e., $\Pr(H_1)$, ranging from 0.01 until 0.99, and for a Bayes factor equal to 4.25, as in Example 3.3.

```
> pi1=seq(0.01,0.99,0.01)
> prior_odds=pi1/(1-pi1)
> BF=4.25
> post_odds=prior_odds*BF
> alpha1=post_odds/(1+post_odds)
```

Fig. 3.2 Posterior probability α_1 of proposition H_1 for values of prior probabilities π_1 ranging from 0.01 to 0.99, and a Bayes factor equal to 4.25 (solid line), 1 (dashed line), and 100 (dotted line)

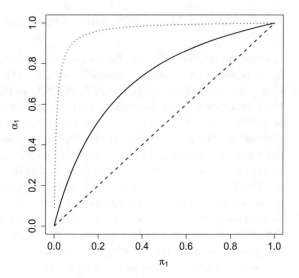

The solid line in Fig. 3.2 shows the value of α_1, the posterior probability of the proposition H_1, as a function of the prior probability, π_1, for BF = 4.25. The plot also shows results for BF = 1 (dashed line) and for BF = 100 (dotted line).

```
> plot(pi1,alpha1,type='l',xlab=expression(pi[1]),
+ ylab=expression(alpha[1]))
> BF=1
> post_odds=prior_odds*BF
> alpha1=post_odds/(1+post_odds)
> lines(pi1,alpha1,lty=2)
> BF=100
> post_odds=prior_odds*BF
> alpha1=post_odds/(1+post_odds)
> lines(pi1,alpha1,lty=3)
```

More generally, it can be observed that the higher the value of the Bayes factor, the smaller the impact of the prior probabilities on posterior probabilities.

3.3 Evidence Evaluation for Continuous Data

The previous section considered the evaluation of scientific evidence as given by discrete data. However, for many types of evidence, measurements result in continuous data.

3.3.1   Normal Model with Known Variance

In some applications, the distribution of measurements exhibits enough regularity
to be captured by standard parametric models, such as the Normal distribution.
One example, introduced earlier in Sect. 1.5.1, is the analysis of magnetism of
black toner on printed documents. Due to the wide distribution and availability
of printing machines, forensic document examiners are commonly requested to
examine documents produced by electrophotographic printing processes that use
dry toner. A question that forensic scientists may be asked to help with is whether
or not two or more documents were printed with the same laser printer. This task
involves the comparison of analytical features of a questioned document with those
of control documents. One such analytical feature is the magnetic flux of toner. It is
thought to be largely influenced by individual settings of the printing device, so that
detectable differences may be expected on documents printed at different instances
using the same or different machines (Biedermann et al., 2016a).

Suspected page substitution is a commonly encountered problem in forensic
document examination. Imagine a case involving a contract consisting of three
pages where the allegation is that the second page has been substituted. It may be of
interest, thus, to investigate the extent to which available measurements of magnetic
flux can be informative in this case.

Consider the following pair of propositions:

H_1 : Page two has been printed by the device used for printing pages one and three
(i.e., the three pages have been printed with the same device).
H_2 : Page two has been printed by a different device.

Denote by $\mathbf{y} = (y_1, \ldots, y_n)$ the measurements of magnetic flux obtained for
the questioned page. Measurements are assumed to be normally distributed with
unknown mean θ and known variance σ^2. The likelihood of the normal random
sample $(y_1, \ldots, y_n)$ can therefore be expressed as

$$f(\mathbf{y} \mid \theta) = \prod_{i=1}^{n} (2\pi\sigma^2)^{-1/2} \exp\left\{ -\frac{1}{2\sigma^2}(y_i - \theta)^2 \right\}. \tag{3.8}$$

It can be shown, (e.g., Bolstad and Curran, 2017), that the likelihood of a normal
random sample is proportional to the likelihood of the sample mean $\bar{y} = \frac{1}{n}\sum_{i=1}^{n} y_i$.
The sample mean is normally distributed with mean θ and variance σ^2/n

$$f(\bar{y} \mid \theta) = (2\pi\sigma^2/n)^{-1/2} \exp\left\{ -\frac{1}{2\sigma^2/n}(\bar{y} - \theta)^2 \right\}. \tag{3.9}$$

In other words, it is possible to reduce the problem to one where a single normal
observation $\bar{y}$ is available.

Next, denote the measurements on uncontested pages by $\{\mathbf{x}_l\} = (x_{lj}, j = 1, \ldots, n$ and $l = 1, 2)$, where the subscript l refers to the page number and j to the number of measurements of magnetic flux obtained for the page l. A normal distribution with mean θ and variance σ^2 is assumed for $\mathbf{x}$, analogously to what has been assumed for $\mathbf{y}$. A conjugate normal prior distribution is chosen for θ, say $\theta \sim N(\mu, \tau^2)$. The Bayes factor can be computed as in (1.16):

$$
\begin{aligned}
\text{BF} &= \frac{f(\bar{y} \mid \mathbf{x}_1, \mathbf{x}_2, H_1)}{f(\bar{y} \mid H_2)} \\
&= \frac{\int f(\bar{y} \mid \theta) f(\theta \mid \mathbf{x}_1, \mathbf{x}_2, H_1) d\theta}{\int f(\bar{y} \mid \theta) f(\theta \mid H_2) d\theta},
\end{aligned}
\tag{3.10}
$$

where $f(\theta \mid \mathbf{x}_1, \mathbf{x}_2, H_1)$ is the posterior distribution of θ, obtained by updating the prior distribution $N(\mu, \tau^2)$ using the measurements $\mathbf{x}_1$ and $\mathbf{x}_2$. This is a normal distribution, $(\theta \mid \mathbf{x}_1, \mathbf{x}_2) \sim N(\mu_x, \tau_x^2)$, with posterior mean μ_x and posterior variance τ_x^2, computed according to the updating rules (2.13) and (2.14). Using the result (1.21), one can easily verify that the density in the numerator is still a normal distribution with mean equal to the posterior mean μ_x and variance equal to the sum of the posterior variance τ_x^2 and the population variance σ^2 divided by the sample size n, i.e., $\tau_x^2 + \sigma^2/n$. In the same way, invoking (1.22), the density in the denominator is still a normal distribution with mean equal to the prior mean μ and variance equal to the sum of the prior variance τ^2 and the population variance σ^2 divided by the sample size n, i.e., $\tau^2 + \sigma^2/n$.

Example 3.5 (Printed Documents) Consider the case described above where a forensic document examiner measures the magnetic flux on two uncontested pages 1 and 3 (Biedermann et al., 2016a). The results are $\mathbf{x}_1 = (16, 15, 15)$ and $\mathbf{x}_2 = (16, 15, 16)$. The measurements for the contested page 2 are $\mathbf{y} = (15, 16, 16)$. Previous experiments allow one to assign the value 0.24 for the population standard deviation σ. Based on the available knowledge regarding the magnetic flux of toner on printed documents, the prior mean μ and the prior variance τ^2 for the unknown quantity of magnetic flux are set equal to 17.5 and 3.92^2, respectively. This means that values of the magnetic flux smaller than 6 and greater than 29 are considered, a priori, to be extremely unlikely.

```
> mu=17.5
> tau2=3.92^2
> sigma2=0.24^2
> x=c(16,15,15,16,15,16)
> y=c(15,16,16)
> nx=length(x)
> ny=length(y)
```

(continued)

Example 3.5 (continued)

The posterior distribution $f(\theta \mid \mathbf{x}_1, \mathbf{x}_2)$ can be obtained by a single application of Bayes theorem with the full set of available measurements $(\mathbf{x}_1, \mathbf{x}_2)$. The posterior parameters μ_x and τ_x^2 can be calculated using the function `post_distr` introduced in Sect. 2.3.1.

```
> mupost=post_distr(sigma2,nx,mean(x),mu,tau2)[1]
> mupost
```

```
[1] 15.50125
```

```
> tau2post=post_distr(sigma2,nx,mean(x),mu,tau2)[2]
> tau2post
```

```
[1] 0.009594006
```

The two marginal densities in the numerator and denominator of the BF in (3.10) can be calculated at the sample mean $\bar{y}$. The exact value of the Bayes factor is given by

```
> BF=dnorm(mean(y),mupost,sqrt(tau2post+sigma2/ny))/
+ dnorm(mean(y),mu,sqrt(tau2+sigma2/ny))
> BF
```

```
[1] 16.03199
```

This value represents moderate support for the proposition of page substitution, compared to the proposition of no page manipulation.

3.3.2 Normal Model with Both Parameters Unknown

So far, the variance of the distribution of the observations has been assumed to be known, though in many practical situations the mean and the variance are both unknown, and it is necessary to choose a prior distribution for the parameter vector (θ, σ^2). The Bayes factor can be computed as in (1.16):

$$
\begin{aligned}
\mathrm{BF} &= \frac{f(\mathbf{y} \mid \mathbf{x}, H_1)}{f(\mathbf{y} \mid H_2)} \\
&= \frac{\int f(\mathbf{y} \mid \theta, \sigma^2) f(\theta, \sigma^2 \mid \mathbf{x}, H_1) \mathrm{d}(\theta, \sigma^2)}{\int f(\mathbf{y} \mid \theta, \sigma^2) f(\theta, \sigma^2 \mid H_2) \mathrm{d}(\theta, \sigma^2)}.
\end{aligned} \tag{3.11}
$$

Consider the case where a conjugate prior distribution for (θ, σ^2) of the form

$$f(\theta, \sigma^2) = f(\theta \mid \sigma^2) f(\sigma^2) \tag{3.12}$$

is chosen. In this distribution, prior beliefs about the population mean θ are calibrated by the scale of measurements of the observations.[1] The conditional distribution $f(\theta \mid \sigma^2)$ is taken to be normal, centered at μ with variance σ^2/n_0, $(\theta \mid \sigma^2) \sim N(\mu, \frac{\sigma^2}{n_0})$. The parameter n_0 can be thought of as the prior sample size for the distribution of θ. As pointed out in Sect. 2.3.1, it formalizes the size of the sample from a normal population that provides an equivalent amount of information about θ. The distribution $f(\sigma^2)$ is taken to be an S times inverse chi-squared distribution with k degrees of freedom, $\sigma^2 \sim S \cdot \chi^{-2}(k)$. It can be shown that this is equivalent to an inverse gamma distribution with shape parameter $\alpha = k/2$ and scale parameter $\beta = S/2$, $\sigma^2 \sim IG(\alpha = k/2, \beta = S/2)$. Alternatively, prior uncertainty about dispersion can be formulated in terms of the precision $\lambda^2 = 1/\sigma^2$. The prior distribution of λ^2 becomes a gamma distribution with shape parameter $\alpha = k/2$ and rate parameter $\beta = S/2$, $\lambda^2 \sim Ga(\alpha = k/2, \beta = S/2)$. For further discussion, see e.g. Bernardo and Smith (2000), Bolstad and Curran (2017) and Robert (2001).

Consider now the posterior distribution of the unknown parameter vector (θ, λ^2) once a vector of observations $\mathbf{x} = (x_1, \ldots, x_n)$ becomes available. It takes the form of a normal–gamma distribution

$$f(\theta, \lambda^2 \mid \mathbf{x}, H_1) = NG(\mu_n, n', \alpha_n, \beta_n),$$

with

$$\mu_n = \frac{n\bar{x} + n_0\mu}{n + n_0} \quad ; \quad n' = n + n_0$$

$$\alpha_n = \alpha + \frac{n}{2};$$

$$\beta_n = \beta + \frac{1}{2}\left[(n-1)s^2 + \frac{n_0 n(\bar{x} - \mu)^2}{n_0 + n}\right],$$

---

[1] Note that in (3.12) population parameters are not, a priori, independent. Whenever this condition is felt to be too restrictive (see, e.g., Robert (2001)), it is also possible to choose a prior distribution as the product of independent priors, $f(\theta, \sigma^2) = f(\theta)f(\sigma^2)$. In this case, the derivation of the posterior distribution can be more demanding.

and $s^2 = \frac{1}{n-1} \sum_{i=1}^{n}(x_i - \bar{x})^2$.

If uncertainty about the two unknown parameters is modeled by means of the conjugate prior distribution in (3.12), the integrations in (3.11) have an analytical solution and the BF can be obtained straightforwardly.

Denote by $\mathbf{y} = (y_1, \ldots, y_{n_y})$ a vector of measurements made on questioned material and consider the sample mean $\bar{y} = \sum_{i=1}^{n_y} y_i$. It can be proved that the marginal density $f(\bar{y} \mid \mathbf{x}, H_1)$ in the numerator is a Student t distribution with $2\alpha + n$ degrees of freedom, centered at μ_n, with spread parameter, denoted s_n, equal to

$$s_n = \frac{n_y(n + n_0)}{n + n_0 + n_y} \left(\alpha + \frac{n}{2} \right) \beta_n^{-1}.$$

This can be denoted as $f_1(\bar{y} \mid \mu_n, s_n, 2\alpha + n)$.

The marginal density $f(y \mid H_2)$ in the denominator is a Student t distribution with k degrees of freedom, centered at μ with spread parameter (precision), denoted s_d, equal to

$$s_d = \frac{n_0 n_y}{n_0 + n_y} \alpha \beta^{-1}$$

(Bernardo and Smith, 2000). This can be denoted as $f_2(\bar{y} \mid \mu, s_d, 2\alpha)$.

The Bayes factor can then be computed as

$$\mathrm{BF} = \frac{f_1(\bar{y} \mid \mu_n, s_n, 2\alpha + n)}{f_2(\bar{y} \mid \mu, s_d, 2\alpha)}. \tag{3.13}$$

Choosing the Parameters of the Normal Prior

The use of a conjugate prior distribution for the mean and the variance of a normal distribution raises the question of how to choose the hyperparameters, as the resulting distribution should suitably reflect available prior knowledge. The prior distribution $f(\theta \mid \sigma^2)$ requires one to choose a value for μ, the measure of location, and a value for n_0. The ratio n_0/n characterizes the relative precision of the prior distribution compared to the precision of the observations. If this ratio is very small, the less informative will be the prior distribution, and the closest will be the posterior distribution to that obtained using a non-informative prior distribution. In fact, when n_0/n approaches zero, the limiting form of the marginal distribution of the population mean θ is $\mathrm{N}(\bar{x}, \sigma^2/n)$, which corresponds to the posterior distribution that would be obtained using a non-informative prior distribution (Robert, 2001). For more specific prior beliefs (i.e., concentrated on a limited range of values), a higher value of n_0 should be chosen.

Regarding the prior distribution of σ^2, consider a number of degrees of freedom $k = 20$ so that the prior mass is distributed rather symmetrically. Suppose also that, based on knowledge available from previous experiments, it is considered that values of σ^2 greater or smaller than 0.05 are equally plausible, so $\Pr(\sigma^2 > 0.05) = 0.5$. The parameter S can be elicited by recalling that $\sigma^2/S \sim \chi^{-2}(k)$ and, analogously, $S \cdot \lambda^2 \sim \chi^2(k)$ so

$$\Pr\left(\sigma^2 > 0.05\right) = \Pr\left(S \cdot \lambda^2 < S \cdot 20\right) = 0.5,$$

where $S \cdot 20$ is the quantile of order 0.5 of a χ^2 distributed random variable with $k = 20$ degrees of freedom.

```
> sigma2=0.05
> k=20
> p=0.5
> q=qchisq(p,k)
> q

[1]  19.33743

> S=q*sigma2
```

Parameter S is then equal to

$$S = 19.3374 \times 0.05 \approx 1.$$

The elicited prior distribution for σ^2 is $IG(\frac{20}{2}, \frac{1}{2})$ and is shown in Fig. 3.3.

Fig. 3.3 Inverse Gamma prior distribution $IG(\frac{20}{2}, \frac{1}{2})$ for σ^2 in Example 3.6

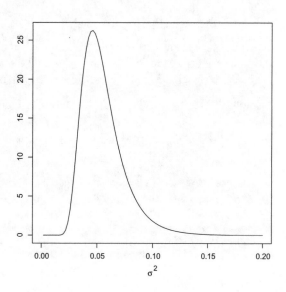

Example 3.6 (Printed Documents—Continued)   Consider again Example 3.5 where magnetic flux was measured on uncontested and questioned pages. The population variance σ^2 was assumed known and equal to 0.0576. Suppose now that a new measuring device is used and that the number of previous experiments (i.e., measurements) conducted with this device is limited. A conjugate prior distribution as in (3.12) is introduced to model prior uncertainty about θ and σ^2.

The prior distribution for $\theta \mid \sigma^2$ can be centered at $\mu = 17.5$ as in Example 3.5 with $n_0 = 0.004$ reflecting a very weak prior belief with respect to the precision of the observations, $\theta \sim N(17.5, \sigma^2/0.004)$.

```
> mu=17.5
> n0=0.004
```

The prior distribution about σ^2 has been elicited above, with $k = 20$ degrees of freedom, and $S = 1$, $\sigma^2 \sim IG(\frac{20}{2}, \frac{1}{2})$, shown in Fig. 3.3.

```
> library(extraDistr)
> S=1
> k=20
> plot(function(x) dinvgamma(x,k/2,S/2),0,0.2,
+ xlab=expression(paste(sigma)^2),ylab='')
```

Note that the function `dinvgamma` is available in the package `extraDistr` (Wolodzko, 2020). Measurements are the same as in Example 3.5.

```
> x=c(16,15,15,16,15,16)
> y=c(15,16,16)
> n=length(x)
> ny=length(y)
```

Let us first consider the marginal density in the numerator of the Bayes factor in (3.13). It is a Student t distribution with $2\alpha + n = k + n = 26$ degrees of freedom, centered at $\mu_n = 15.5$ with spread parameter $s_n = 20.6724$.

```
> mun=(n*mean(x)+n0*mu)/(n+n0)
> mun

[1] 15.50133

> s2=sum((x-mean(x))^2)
> bn=S/2+(s2+n0*n*(mean(x)-mu)^2*(n0+n)^(-1))/2
> sn=ny*(n+n0)/(n+n0+ny)*(k+n)/2*bn^(-1)
> sn

[1] 20.6724
```

(continued)

Example 3.6 (continued)
The marginal density at the denominator of the Bayes factor in (3.13) is a
Student t distribution with $2\alpha = k = 20$ degrees of freedom, centered at
$\mu = 17.5$ with spread parameter $s_d = 0.0799$.

```
> sd=ny*n0/(n0+ny)*k/S
> sd
```

```
[1] 0.07989348
```

The density of a non-central Student t distributed random variable can be calculated using the function dstp, available in the package LaplacesDemon
(Hall et al., 2020). The Bayes factor can be obtained as

```
> library(LaplacesDemon)
> BF=dstp(mean(y),mun,sn,k+n)/dstp(mean(y),mu,sd,k)
> BF
```

```
[1] 13.88188
```

The Bayes factor represents moderate support for the proposition according
to which page two has been printed by the same device as the one used for
printing pages one and three, compared to the proposition according to which
page two has been printed by a different device.

It is worth emphasizing that the BF is highly sensitive to the choice of the prior
(see Sect. 1.11). A sensitivity analysis should therefore be conducted.

3.3.3 Normal Model for Inference of Source

Consider again a case as described in Sect. 3.3.1, involving the analysis of toner on
printed documents. Magnetic flux was considered as a feature of interest because it
is largely influenced by the settings of the printing device. Suppose now that more
than one potential source (i.e., printing device) is available for examination. The
issue of interest is which of two machines has been used to print a questioned
document (e.g., a contested contract). The propositions of interest can be defined
as follows:

H_1 : The questioned document has been printed with machine A.
H_2 : The questioned document has been printed with machine B.

The two potential sources, i.e., machines A and B, are used to print documents
under controlled conditions. The measurements made on documents printed by
the two devices are denoted $\{\mathbf{x}_p\} = (\mathbf{x}_{pi}, \ p = A, B \text{ and } i = 1, \ldots, m)$, with

$\mathbf{x}_{pi} = (x_{pi1}, \ldots, x_{pin})$ denoting the vector of n measurements for each analyzed page, $i = 1, \ldots, m$, from each printer $p = A, B$. Measurements are assumed to be normally distributed with unknown mean θ_p, $p = A, B$, and variance σ^2. The variance is assumed to be known and equal for the two devices. A conjugate normal prior distribution is taken for the unknown mean θ_p, say $\theta_p \sim N(\mu_p, \tau_p^2)$, $p = A, B$.

Measurements on the questioned document are denoted by $\mathbf{y} = (\mathbf{y}_1, \ldots, \mathbf{y}_q)$, with $\mathbf{y}_j = (y_{j1}, \ldots, y_{jn})$ denoting the vector of n measurements from each contested page $j = 1, \ldots, q$. For cases in which $q > 1$, it is assumed that all pages have been printed with a single device. The distribution of measurements on the questioned document is also taken to be normal. The sample mean $\bar{y} = \frac{1}{nq} \sum_{j=1}^{q} \sum_{k=1}^{n} y_{jk}$ has a normal distribution with mean θ_p and variance σ^2/nq, $(\bar{Y} \mid \theta_p, \sigma^2/nq) \sim N(\theta_p, \sigma^2/nq)$.

The Bayes factor can be computed as

$$\begin{aligned} \text{BF} &= \frac{\int f(\bar{y} \mid \theta_A) f(\theta_A \mid \mathbf{x}_A) d\theta_A}{\int f(\bar{y} \mid \theta_B) f(\theta_B \mid \mathbf{x}_B) d\theta_B} \\ &= \frac{f(\bar{y} \mid \mathbf{x}_A, H_1)}{f(\bar{y} \mid \mathbf{x}_B, H_2)}. \end{aligned} \tag{3.14}$$

The marginal probability density in the numerator can be obtained in closed form. It is a normal distribution with mean equal to the posterior mean $\mu_{A,x}$ and variance equal to the sum of the posterior variance $\tau_{A,x}^2$ and population variance σ_A^2/nq (where nq is the total number of observations), that is, $f(\bar{y} \mid \mathbf{x}_A, H_1) = N(\mu_{A,x}, \tau_{A,x}^2 + \sigma^2/nq)$. In the same way, one can obtain the marginal probability density in the denominator, $f(\bar{y} \mid \mathbf{x}_B, H_2) = N(\mu_{B,x}, \tau_{B,x}^2 + \sigma^2/nq)$. As observed in Sect. 3.3.1, the numerator and the denominator of (3.14) can be calculated as the densities of two normally distributed random variables, $N(\mu_{A,x}, \tau_{A,x}^2 + \sigma^2/nq)$ and $N(\mu_{B,x}, \tau_{B,x}^2 + \sigma^2/nq)$, at the sample mean $\bar{y}$ of the measurements on the questioned document.

Example 3.7 (Printed Documents)   Consider a type of case and propositions as introduced above, and suppose that there is only one contested page, that is, $q = 1$. Measurements of the magnetic flux lead to the following results: $\mathbf{y} = (20, 20, 21)$ (i.e., $n = 3$ measurements are taken). Two pages are printed with each printing device. The results are as follows (Biedermann et al., 2016a):

	Printer A	Printer B
Page 1	20 20 19	21 20 21
Page 2	20 21 20	21 22 21

(continued)

Example 3.7 (continued)

The available data thus are

```
> xa=c(20,20,19,20,21,20)
> xb=c(21,20,21,21,22,21)
> y=c(20,20,21)
> n=length(y)
```

The population standard deviation σ is taken to be equal to 0.24, as in Example 3.5. We also choose the same prior distribution as used in Example 3.5 to describe uncertainty about the magnetic flux of toner printed by the two printing devices. Thus, $\mu_A = \mu_B = 17.5$ and $\tau_A^2 = \tau_B^2 = 3.92^2$.

```
> sigma2=0.24^2
> na=length(xa)
> nb=length(xb)
> mu=17.5
> tau2=3.92^2
```

The posterior distributions $f(\theta_A \mid \mathbf{x}_A)$ and $f(\theta_B \mid \mathbf{x}_B)$ can be obtained by a single application of Bayes theorem using the full set of available measurements for each printer. The posterior parameters $\mu_{A,x}$, $\mu_{B,x}$, $\tau_{A,x}^2$ and $\tau_{B,x}^2$ can be calculated using the function post_distr:

```
> muapost=post_distr(sigma2,na,mean(xa),mu,tau2)[1]
> tauapost=post_distr(sigma2,na,mean(xa),mu,tau2)[2]
> mubpost=post_distr(sigma2,nb,mean(xb),mu,tau2)[1]
> taubpost=post_distr(sigma2,nb,mean(xb),mu,tau2)[2]
```

The two marginal densities in the numerator and denominator of the BF in (3.14) can be calculated at the observed value $\bar{y}$. The BF can thus be computed as the ratio of two marginal densities:

```
> BF=dnorm(mean(y),muapost,sqrt(sigma2/n+tauapost))/
+ dnorm(mean(y),mubpost,sqrt(sigma2/n+taubpost))
> BF
```

```
[1] 304.7886
```

This value represents moderately strong support for the proposition according to which the questioned page been printed using device *A*, rather than using device *B*.

Consider a "$0 - l_i$" loss function as in Table 1.4. The optimal decision is to accept the view according to which the questioned page was printed by the device *A* (as stated by proposition H_1), rather than by device *B*, whenever

$$BF > \frac{l_1/l_2}{\pi_1/\pi_2}.$$

If the odds are evens, and a symmetric loss function is felt to be appropriate, the Bayes decision is to accept the view according to which the questioned document has been printed with machine A (B) whenever the BF is greater (smaller) than 1.

When available information is limited, one may choose a non-informative prior distribution for (θ, σ^2) that can be specified as

$$f(\theta, \sigma^2) = \frac{1}{\sigma^2}. \tag{3.15}$$

In this case, the marginal distribution in the numerator of the BF is proportional to a Student t distribution with $n_A - 1$ degrees of freedom, centered at the sample mean $\bar{x}_A$ with spread parameter s_n equal to

$$s_n = \frac{n_A n q}{(n_A + n q) s_A^2},$$

where $s_A = \frac{1}{n_A-1} \sum_{i=1}^{n_A} (x_A - \bar{x}_A)^2$, n_A is the total number of observations from device A, and nq is the total number of measurements from the q contested pages (i.e., n measurements for each contested page). This can be denoted as $f_1(\bar{y} \mid \bar{x}_A, s_n, n_A - 1)$.

Vice versa, the marginal distribution in the denominator of the BF is proportional to a Student t distribution with $n_B - 1$ degrees of freedom, centered at the sample mean $\bar{x}_B$ with spread parameter s_d equal to

$$s_d = \frac{n_B n q}{(n_B + n q) s_B^2},$$

where $s_B = \frac{1}{n_B-1} \sum_{i=1}^{n_B} (x_B - \bar{x}_B)^2$ and n_B is the total number of observations from device B. This can be denoted as $f_2(\bar{y} \mid \bar{x}_B, s_d, n_B - 1)$.

The Bayes factor can then be obtained as

$$BF = \frac{f_1(\bar{y} \mid \bar{x}_A, s_n, n_A - 1)}{f_2(\bar{y} \mid \bar{x}_B, s_d, n_B - 1)}. \tag{3.16}$$

Example 3.8 (Printed Documents—Continued) In Example 3.7, a normal prior distribution has been used for (θ, σ^2). Consider now a non-informative prior distribution as in (3.15). In order to compute the Bayes factor, one must first obtain the spread parameters s_n and s_d under the competing propositions.

(continued)

Example 3.8 (continued)
```
> s2a=var(xa)
> sn=na*n/((na+n)*s2a)
> s2b=var(xb)
> sd=nb*n/((nb+n)*s2b)
```

Note that in this case the number of contested pages q is set equal to 1. The density of a non-central Student t distributed random variable can be obtained using the function `dstp` available in the package `LaplacesDemon` (Hall et al., 2020). The Bayes factor can be obtained as follows:

```
> library(LaplacesDemon)
> BF=dstp(mean(y),mean(xa),sn,na-1)/
+ dstp(mean(y),mean(xb),sd,nb-1)
> BF
```

```
[1] 2.197
```

The Bayes factor represents weak support for the proposition according to which the questioned document has been printed with machine A, rather than with machine B.

More Than Two Propositions

Consider now the case where more than two devices are available. As in Sect. 1.6, the question is how to evaluate measurements made on questioned and known items (i.e., documents), as the BF involves pairwise comparisons. A scaled version of the marginal likelihood may be reported as in (1.27).

Example 3.9 (Printed Documents, More Than Two Propositions) Recall Example 3.7, and assume that a third printer, machine C, is available for comparative examinations. The propositions of interest are therefore:

H_1 : The questioned document has been printed with machine A.
H_2 : The questioned document has been printed with machine B.
H_3 : The questioned document has been printed with machine C.

Two pages are printed with the additional printing device C. All results, including those from machines A and B, are as follows:

(continued)

Example 3.9 (continued)

	Printer A	Printer B	Printer C
Page 1	20 20 19	21 20 21	21 20 21
Page 2	20 21 20	21 22 21	20 21 20

Let the prior distribution describing uncertainty about the magnetic flux characterizing machine C be the same as introduced previously, that is $\mu_C = 17.5$ and $\tau_C^2 = 3.92^2$. First, the posterior distribution $f(\theta_C \mid \mathbf{x}_C)$ is calculated:

```
> xc=c(21,20,21,20,21,20)
> nc=length(xc)
> mucpost=post_distr(sigma2,nc,mean(xc),mu,tau2)[1]
> taucpost=post_distr(sigma2,nc,mean(xc),mu,tau2)[2]
```

Next, consider the marginal likelihoods of the sample mean that can be obtained as

```
> mla=dnorm(mean(y),muapost,sqrt(sigma2/n+tauapost))
> mlb=dnorm(mean(y),mubpost,sqrt(sigma2/n+taubpost))
> mlc=dnorm(mean(y),mucpost,sqrt(sigma2/n+taucpost))
```

The scaled version of the marginal likelihoods then is

```
> smla=mla/(mla+mlb+mlc)
> smlb=mlb/(mla+mlb+mlc)
> smlc=mlc/(mla+mlb+mlc)
> round(c(smla,smlb,smlc),5)

[1] 0.18593 0.00061 0.81346
```

Recall from Sect. 1.6 that this is equivalent to reporting the posterior probability of competing propositions with equal prior probabilities. Therefore, if $\Pr(H_1) = \Pr(H_2) = \Pr(H_3) = \frac{1}{3}$, then proposition H_3 has received the greatest evidential support.

Alternatively, the analyst may also consider the possibility of aggregating propositions H_1 and H_2 and consider:

H_1 : The questioned document has been printed with machine C.
$\bar{H}_1$: The questioned document has been printed with machine A or B.

Example 3.10 (Printed Documents, More Than Two Propositions—Continued) When considering a single proposition H_1 compared to a composite proposition $\bar{H}_1$ as defined above, the Bayes factor can be obtained as in (1.28), with $\Pr(H_1) = 1/3$ and $\Pr(\bar{H}_1) = 2/3$.

```
> p=1/3
> mlc*(1-p)/(mla*p+mlb*p)

[1] 8.72179
```

3.3.4 Score-Based Bayes Factor

As mentioned previously in Sect. 1.5.2, it may not be possible to specify a probability model for some types of forensic evidence and data. An example was given in Sect. 3.2.3 for discrete data regarding consecutive matching striations, used to quantify the extent of agreement between marks on bullets.

Consider now a case where a saliva trace is collected at the crime scene. The salivary microbiome is analyzed as well as that of traces originating from a known source, Mr. X, with the aim of discriminating between the following competing propositions:

H_1 : The saliva trace comes from Mr. X.
H_2 : The saliva trace comes from the twin brother of Mr. X.

Note that the proposition H_2 represents an extreme case of relatedness. To investigate this type of case, consider the data collected by Scherz (2021). This longitudinal study involving 30 monozygotic twins has shown the potential of salivary microbiome profiles to discriminate between closely related individuals (Scherz et al., 2021). This may represent an alternative method when standard DNA profiling analyses yield no useful results.

In the study by Scherz (2021), four salivary samples have been collected from each participant. The first at the beginning of the study, and the others after 1, 12, and 13 months. Given the complex composition of microbiota, a distance can be calculated to compare microbiota profiles. One possibility is the Jaccard distance, obtained by dividing the number of amplicon sequence variants (AVSs) shared by the two samples by the number of distinct AVSs in the two compared samples. This measure has shown good discriminatory power. Other distances (e.g., Jensen–Shannon) can be calculated (Scherz, 2021).

The intra-individual variability was studied by comparing all four samples of each individual. The intra-pair variability was evaluated by comparing pairs of samples from related individuals (here: homozygous twins). The inter-individual variability was studied by comparing samples of unrelated individuals (Fig. 3.4).

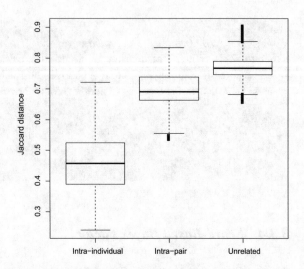

Fig. 3.4 Jaccard distances for salivary microbiota compositions of pairs of samples from individual persons (intra-individual), pairs of related persons (intra-pair), and pairs of unrelated persons (unrelated) [Source of data: (Scherz et al., 2021)]

Let $\delta(y, x)$ denote the distance between the analytical features of questioned material (i.e., a saliva trace of unknown origin) and control material (i.e., a saliva sample from Mr. X). A score-based Bayes factor (sBF) can be defined as follows:

$$\text{sBF} = \frac{g(\delta(x, y) \mid H_1)}{g(\delta(x, y) \mid H_2)}. \tag{3.17}$$

To obtain a value for this sBF, it is necessary to study the probability distribution of the calculated score under the competing propositions. However, the limited number of samples per individual, available for pairwise comparison, might make it difficult to assess the numerator, which is specific for a given person of interest. To address this problem, Davis et al. (2012) propose the use of a database of simulated samples to help with the construction of probability distributions for scores.

In the example studied here, a maximum number of 6 intra-volunteer comparisons are available for each participant. A viable alternative is to perform a so-called common-source comparison,[2] and use the limited number of items from all participants, provided that one is willing to assume a generic probability distribution for all individuals in the numerator. In the same way, a generic probability distribution is used at the denominator in all cases where a twin is assumed as the alternative source of the salivary trace (Bozza et al., 2022).

Denote by $\{Z_{ij}^1, i = 1, \ldots, m_1, j = 1, \ldots, n_1\}$ the intra-individual distances and by $\{Z_{ij}^2, i = 1, \ldots, m_2, j = 1, \ldots, n_2\}$ the intra-pair distances, where m_1 (m_2) are the number of distinct individuals (couples of twin brothers) and n_1 (n_2) are the number of distances calculated for each individual (couple). A normal distribution is used for both the numerator and denominator to model the *within-source* variation

---

[2] See Sect. 1.5.2 on the difference between specific-source and common-source propositions.

(i.e., the variation between distances characterizing materials originating from the same individual and from the same couple of twins, respectively), $Z_{ij}^p \sim N(\theta_p, \sigma_p^2)$, where $p = \{1, 2\}$. Different distributions can be used to describe the between-source variation (i.e., the variation between distances characterizing materials originating from different individuals and from different couples of twins, respectively). Here, a normal distribution is retained, $\theta_p \sim N(\mu_p, \tau_p^2)$. The mean vector between sources μ_p, the within-source variance σ_p^2, and the between-source variance τ_p^2 can be estimated from the background data:

$$\hat{\mu}_p = \bar{z}_p = \frac{1}{m_p n_p} \sum_{i=1}^{m_p} \sum_{j=1}^{n_p} z_{ij}^p \tag{3.18}$$

$$\hat{\sigma}_p^2 = \frac{1}{m_p(n_p - 1)} \sum_{i=1}^{m_p} \sum_{j=1}^{n_p} (z_{ij}^p - \bar{z}_i)^2 \tag{3.19}$$

$$\hat{\tau}_p^2 = \frac{1}{m_p - 1} \sum_{i=1}^{m_p} (\bar{z}_i^p - \bar{z}_p)^2 - \frac{\hat{\sigma}_p^2}{n_p}, \tag{3.20}$$

where $\bar{z}_i^p = \sum_{j=1}^{n_p} z_{ij}$.

Example 3.11 (Saliva Traces)   Consider a case where a saliva trace is recovered at a crime scene and a sample is taken from a person of interest for comparative purposes. The Jaccard distance between the microbiota composition of recovered and control sample is equal to 0.51.

```
> d=0.51
```

The propositions are H_1, the compared items come from the same source, and H_2, the compared items come from different sources (twins). Suppose that the estimated means between sources in (3.18) are 0.454 and 0.769; the estimated within-source variances in (3.19) are 0.0057 and 0.00067; the estimated between-source variances in (3.20) are 0.0028 and 0.0024 (Source of data: Scherz (2021)).

```
> mu1=0.454
> mu2=0.769
> sigma1=0.0057
> sigma2=0.00067
> tau1=0.0028
> tau2=0.0024
```

The Bayes factor can then be obtained straightforwardly as in (3.17)

(continued)

Example 3.11 (continued)
```
> BF=dnorm(d,mu1,sqrt(tau1+sigma1))/
+ dnorm(d,mu2,sqrt(tau2+sigma2))
> BF

[1] 27766.33
```
The Bayes factor provides very strong support for the proposition that the saliva traces originate from the same individual rather than from two different individuals (twins).

Note that a higher value of the BF is expected whenever the alternative proposition H_2 involves unrelated individuals. The inspection of Fig. 3.4 highlights that higher distances are recorded in this type of case.

The between-source variability can also be modeled by a kernel density distribution, as presented in Bozza et al. (2022). See also Sect. 3.4.1.2, where a detailed description of the kernel density approach is given for two-level multivariate data.

3.4   Multivariate Data

Forensic scientists encounter multivariate data in contexts where the examined objects and materials can be described by several variables. Examples are glass fragments that are searched and recovered on the clothing of a person of interest and on a crime scene, or seized materials supposed to contain illicit substances. Such materials may be analyzed and compared on the basis of their chemical compounds as well as their physical characteristics. Multivariate data also arise in other forensic science disciplines, such as handwriting examination. Handwritten characters can, in fact, be described by means of several variables, such as the width, the height, the surface, the orientation of the strokes, or by Fourier descriptors (Marquis et al., 2005). In addition, an emerging topic that forensic document examiners nowadays encounter is handwriting (e.g., signatures) on digital tablets. Such electronic devices provide several static (e.g., length of a signature) and dynamic features (e.g., speed) that can be used as variables to describe signatures (Linden et al., 2018). These developments have led to substantial databases that often present a complex dependence structure, a large number of variables, and multiple sources of variation.

3.4.1 Two-Level Models

Denote by p the number of characteristics (variables) observed on items of a particular evidential type. Suppose that continuous measurements of these variables are available on a random sample of m sources with n items from each source. For handwriting evidence, a source is a single writer, with n characters from each writer and p observed characteristics that pertain to the shape of handwritten characters. For glass evidence, a source is a window, with n replicate measurements from a glass fragment originating from each window and p observed characteristics given by concentrations in elemental composition. The background data can be denoted by $\mathbf{z}_{ij} = (z_{ij1}, \ldots, z_{ijp})$, where $i = 1, \ldots, m$ denotes the number of sources (e.g., windows), $j = 1, \ldots, n$ denotes the number of items for each source (e.g., replicate measurements from a glass fragment), and p is the number of variables.

This data structure suggests a two-level hierarchy, accounting for two sources of variation: the variation between replicate measurements within the same source (the so-called within-source variation) and the variation between sources (the so-called between-source variation).

3.4.1.1 Normal Distribution for the Between-Source Variability

In some applications, data exhibit regularity that can reasonably be described using standard probabilistic models. For example, the within-source variability and the between-source variability may be modeled by a normal distribution. A Bayesian statistical model for the evaluation of trace evidence for two-level normally distributed multivariate data was proposed by Aitken and Lucy (2004) in the context of evaluating the elemental composition of glass fragments. To illustrate this model, denote the mean vector within source i by $\boldsymbol{\theta}_i$. Denote by W the matrix of within-source variances and covariances. The distribution of Z_{ij} for the within-source variation is taken to be normal, $Z_{ij} \sim N(\boldsymbol{\theta}_i, W)$. For the between-source variation, the mean vector between sources is denoted by $\boldsymbol{\mu}$, and the matrix of between-source variances and covariances by B. The distribution of the $\boldsymbol{\theta}_i$ is taken to be normal, $\boldsymbol{\theta}_i \sim N(\boldsymbol{\mu}, B)$.

Measurements are available on items from an unknown source (recovered material) as well as measurements on items from a known source (control material). The examined items may or may not come from the same source. Competing propositions may be formulated as follows:

H_1 : The recovered and the control item originate from the same source.
H_2 : The recovered and the control item originate from different sources.

Denote the measurements on recovered and control items by, respectively, $\mathbf{y} = (\mathbf{y}_1, \ldots, \mathbf{y}_{n_y})$ and $\mathbf{x} = (\mathbf{x}_1, \ldots, \mathbf{x}_{n_x})$, where $\mathbf{y}_j = (y_{j1}, \ldots, y_{jp})$, $\mathbf{x}_j = (x_{j1}, \ldots, x_{jp})$, $j = 1, \ldots, n_{y(x)}$. A Bayes factor can be derived as in (1.15):

$$\text{BF} = \frac{f(\mathbf{y}, \mathbf{x} \mid H_1)}{f(\mathbf{y}, \mathbf{x} \mid H_2)}. \tag{3.21}$$

The distribution of the measurements on the recovered and control materials is taken to be normal, with vector means $\boldsymbol{\theta}_y$ and $\boldsymbol{\theta}_x$, and covariance matrices W_y and W_x. Thus,

$$(Y \mid \boldsymbol{\theta}_y, W_y) \sim \text{N}(\boldsymbol{\theta}_y, W_y) \qquad ; \qquad (X \mid \boldsymbol{\theta}_x, W_x) \sim \text{N}(\boldsymbol{\theta}_x, W_x). \tag{3.22}$$

The Bayes factor is the ratio of two probability densities of the form $f(\mathbf{y}, \mathbf{x} \mid H_i) = f_i(\mathbf{y}, \mathbf{x} \mid \boldsymbol{\mu}, W, B), i = 1, 2$. The probability density in the numerator is given by

$$f_1(\mathbf{y}, \mathbf{x} \mid \boldsymbol{\mu}, W, B) = \int_{\theta} f(\mathbf{y} \mid \boldsymbol{\theta}, W) f(\mathbf{x} \mid \boldsymbol{\theta}, W) f(\boldsymbol{\theta} \mid \boldsymbol{\mu}, B) d\boldsymbol{\theta}, \tag{3.23}$$

where

$$f(\mathbf{y} \mid \boldsymbol{\theta}, W) = |2\pi|^{-pn_y/2} |W|^{-n_y/2} \exp\left[-\frac{1}{2} \sum_{j=1}^{n_y} (\mathbf{y}_j - \boldsymbol{\theta})' W^{-1} (\mathbf{y}_j - \boldsymbol{\theta}) \right], \tag{3.24}$$

$f(\mathbf{x} \mid \boldsymbol{\theta}, W)$ has the same probabilistic structure as $f(\mathbf{y} \mid \boldsymbol{\theta}, W)$, and

$$f(\boldsymbol{\theta} \mid \boldsymbol{\mu}, B) = |2\pi|^{-p/2} |B|^{-1/2} \exp\left[-\frac{1}{2} (\boldsymbol{\theta} - \boldsymbol{\mu})' B^{-1} (\boldsymbol{\theta} - \boldsymbol{\mu}) \right]. \tag{3.25}$$

In the denominator, where $\mathbf{y}$ and $\mathbf{x}$ are taken to be independent, the probability density is given by

$$f_2(\mathbf{y}, \mathbf{x} \mid \boldsymbol{\mu}, W, B) = f_2(\mathbf{y} \mid \boldsymbol{\theta}, W, B) \times f_2(\mathbf{x} \mid \boldsymbol{\theta}, W, B) \tag{3.26}$$

$$= \int_{\theta} f(\mathbf{y} \mid \boldsymbol{\theta}, W) f(\boldsymbol{\theta} \mid \boldsymbol{\mu}, B) d\boldsymbol{\theta} \int_{\theta} f(\mathbf{x} \mid \boldsymbol{\theta}, W) f(\boldsymbol{\theta} \mid \boldsymbol{\mu}, B) d\boldsymbol{\theta}.$$

This is equivalent to the algebraic expression of the Bayes factor in (1.23). In the numerator, under proposition H_1, the source means $\boldsymbol{\theta}_y$ and $\boldsymbol{\theta}_x$ are assumed equal, say $\boldsymbol{\theta}_y = \boldsymbol{\theta}_x = \boldsymbol{\theta}$. In the denominator, under proposition H_2, the source means $\boldsymbol{\theta}_y$ and $\boldsymbol{\theta}_x$ are assumed to be different.

The integrals in (3.23) and (3.26) have an analytical solution. A proof is given by Aitken and Lucy (2004). The numerator can be shown to be equal to

$$f(\mathbf{y}, \mathbf{x} \mid H_1) = |2\pi W|^{-(n_y+n_x)/2} |2\pi B|^{-1/2} |2\pi \left[(n_y + n_x)W^{-1} + B^{-1} \right]^{-1} |^{\frac{1}{2}}$$

$$\times \exp\left\{ -\frac{1}{2} \left[F_1 + F_2 + \text{tr}\left(S_y W^{-1} \right) + \text{tr}\left(S_x W^{-1} \right) \right] \right\}, \tag{3.27}$$

where:

$$F_1 = (\bar{\mathbf{w}} - \boldsymbol{\mu})' \left(\frac{W}{n_y + n_x} + B \right)^{-1} (\bar{\mathbf{w}} - \boldsymbol{\mu}),$$

$$F_2 = (\bar{\mathbf{y}} - \bar{\mathbf{x}})' \left(\frac{W}{n_y} + \frac{W}{n_x} \right)^{-1} (\bar{\mathbf{y}} - \bar{\mathbf{x}}),$$

$$\bar{\mathbf{w}} = \frac{1}{n_y + n_x} \left(\sum_{j=1}^{n_y} \mathbf{y}_j + \sum_{j=1}^{n_x} \mathbf{x}_j \right), \bar{\mathbf{y}} = \frac{1}{n_y} \sum_{j=1}^{n_y} \mathbf{y}_j \text{ and } \bar{\mathbf{x}} = \frac{1}{n_x} \sum_{j=1}^{n_x} \mathbf{x}_j,$$

$$S_y = \sum_{j=1}^{n_y} (\mathbf{y}_j - \bar{\mathbf{y}}) (\mathbf{y}_j - \bar{\mathbf{y}})', S_x = \sum_{j=1}^{n_x} (\mathbf{x}_j - \bar{\mathbf{x}}) (\mathbf{x}_j - \bar{\mathbf{x}})'.$$

Consider the first factor in the denominator, $f_2(\mathbf{y} \mid \boldsymbol{\theta}, W, B)$. It can be obtained as

$$f_2(\mathbf{y} \mid \boldsymbol{\mu}, W, B) = \mid 2\pi W \mid^{-n_y/2} \mid 2\pi B \mid^{-1/2} \mid 2\pi (n_y W^{-1} + B^{-1})^{-1} \mid^{1/2}$$

$$\times \exp \left\{ -\frac{1}{2} \left[(\bar{\mathbf{y}} - \boldsymbol{\mu})'(n_y^{-1} W + B)^{-1}(\bar{\mathbf{y}} - \boldsymbol{\mu}) + \mathrm{tr} \left(S_y W^{-1} \right) \right] \right\}.$$

$$(3.28)$$

The second factor $f_2(\mathbf{x} \mid \boldsymbol{\theta}, W, B)$ can be obtained analogously as

$$f_2(\mathbf{x} \mid \boldsymbol{\mu}, W, B) = \mid 2\pi W \mid^{-n_x/2} \mid 2\pi B \mid^{-1/2} \mid 2\pi (n_x W^{-1} + B^{-1})^{-1} \mid^{1/2}$$

$$\times \exp \left\{ -\frac{1}{2} \left[(\bar{\mathbf{x}} - \boldsymbol{\mu})'(n_x^{-1} W + B)^{-1}(\bar{\mathbf{x}} - \boldsymbol{\mu}) + \mathrm{tr} \left(S_x W^{-1} \right) \right] \right\}.$$

$$(3.29)$$

The Bayes factor in (3.21) then is the ratio between (3.27) and the product between (3.28) and (3.29), respectively. After some manipulation, the BF can be obtained as the ratio between

$$\mid 2\pi \left[(n_y + n_x) W^{-1} + B^{-1} \right]^{-1} \mid^{1/2} \exp \left\{ -\frac{1}{2} (F_1 + F_2) \right\} \qquad (3.30)$$

and

$$\mid 2\pi B \mid^{-1/2} \mid 2\pi (n_y W^{-1} + B^{-1})^{-1} \mid^{1/2} \mid 2\pi (n_x W^{-1} + B^{-1})^{-1} \mid^{1/2}$$

$$\times \exp \left\{ -\frac{1}{2} (F_3 + F_4) \right\}, \qquad (3.31)$$

where:

$$F_3 = (\boldsymbol{\mu} - \boldsymbol{\mu}^*)' \left\{ \left(\frac{W}{n_y} + B \right)^{-1} + \left(\frac{W}{n_x} + B \right)^{-1} \right\} (\boldsymbol{\mu} - \boldsymbol{\mu}^*),$$

$$F_4 = (\bar{\mathbf{y}} - \bar{\mathbf{x}})' \left(\frac{W}{n_y} + \frac{W}{n_x} + 2B \right)^{-1} (\bar{\mathbf{y}} - \bar{\mathbf{x}}),$$

$$\boldsymbol{\mu}^* = \left\{ \left(\frac{W}{n_y} + B \right)^{-1} + \left(\frac{W}{n_x} + B \right)^{-1} \right\}^{-1} \times \left\{ \left(\frac{W}{n_y} + B \right)^{-1} \bar{\mathbf{y}} + \left(\frac{W}{n_x} + B \right)^{-1} \bar{\mathbf{x}} \right\}.$$

The mean vector between sources μ, the within-source covariance matrix W, and the between-source covariance matrix B can be estimated using the available background data:

$$\hat{\mu} = \bar{\mathbf{z}} = \frac{1}{mn} \sum_{i=1}^{m} \sum_{j=1}^{n} \mathbf{z}_{ij}, \tag{3.32}$$

$$\hat{W} = \frac{1}{m(n-1)} \sum_{i=1}^{m} \sum_{j=1}^{n} (\mathbf{z}_{ij} - \bar{\mathbf{z}}_i)(\mathbf{z}_{ij} - \bar{\mathbf{z}}_i)', \tag{3.33}$$

$$\hat{B} = \frac{1}{m-1} \sum_{i=1}^{m} (\bar{\mathbf{z}}_i - \bar{\mathbf{z}})(\bar{\mathbf{z}}_i - \bar{\mathbf{z}})' - \frac{\hat{W}}{n}, \tag{3.34}$$

where $\bar{\mathbf{z}}_i = \frac{1}{n} \sum_{j=1}^{n} \mathbf{z}_{ij}$.

Example 3.12 (Glass Evidence) Consider a case in which two glass fragments are recovered on the jacket of an individual who is suspected to be involved in a crime. Two glass fragments are collected at the crime scene for comparative purposes. The competing propositions are:

H_1 : The recovered and known glass fragments originate from the same source (broken window at the crime scene).

H_2 : The recovered and known glass fragments originate from different sources.

For each fragment, three variables are considered: the logarithmic transformation of the ratios Ca/K, Ca/Si, and Ca/Fe (Aitken and Lucy, 2004). Two replicate measurements are available for each fragment. Measurements on the two recovered fragments are

$$\mathbf{y}_1 = \begin{pmatrix} 3.77379 \\ -0.89063 \\ 2.62038 \end{pmatrix}, \ \mathbf{y}_2 = \begin{pmatrix} 3.93937 \\ -0.89343 \\ 2.63860 \end{pmatrix}.$$

Measurements on the two control fragments are

$$\mathbf{x}_1 = \begin{pmatrix} 3.84396 \\ -0.91010 \\ 2.65437 \end{pmatrix}, \ \mathbf{x}_2 = \begin{pmatrix} 3.72493 \\ -0.89811 \\ 2.61933 \end{pmatrix}.$$

Consider the database named `glass-data.txt`. This database is part of the supplementary material of Aitken and Lucy (2004) and contains $n = 5$ replicate measurements of the elemental concentration of glass fragments

(continued)

Example 3.12 (continued)

from several windows ($m = 62$). The variables of interest (i.e., the logarithmic transformation of the ratios Ca/K, Ca/Si, and Ca/Fe) are displayed in columns 6, 7 and 8, while the object (window) identifier is in column 9.

```
> population=read.table("glass-data.txt", header=T)
> variables=c(6,7,8)
> grouping.item=9
```

Measurements from the recovered fragments, $\mathbf{y} = (\mathbf{y}_1, \mathbf{y}_2)$, and measurements from the control fragments, $\mathbf{x} = (\mathbf{x}_1, \mathbf{x}_2)$, were selected from the available replicate measurements for the first group (window). The first two replicate measurements were selected to act as recovered data, while the last two replicate measurements were selected to act as control data

```
> item=1
> recovered=population[which(population[,grouping.
+ item]==item),][1:2,variables]
> recovered

  logCaK  logCaSi logCaFe
1 3.77379 -0.89063 2.62038
2 3.93937 -0.89343 2.63860

> control=population[which(population[,grouping.
+ item]==item),][4:5,variables]
> control

  logCaK  logCaSi logCaFe
4 3.72493 -0.89811 2.61933
5 3.66573 -0.89693 2.76393
```

Data concerning measurements from the first window were then excluded from the database

```
> pop.back <- population[-which(population[,grouping.
+ item]==item),]
```

The database named pop.back will serve as background data and can be used to estimate the model parameters μ, W and B as in (3.32), (3.33), and (3.34) by means of the function two.level.mv.WB contained in the routines file two_level_functions.r. This file is part of the supplementary materials available on the website of this book (on

(continued)

Example 3.12 (continued)
`http://link.springer.com/)` and can be run in the R console by inserting the command

```
> source('two_level_functions.r')
```

The mean vector between sources, the within-source covariance matrix, and the between-source covariance matrix can therefore be obtained as follows:

```
> WB <- two.level.mv.WB(pop.back,variables,
+ grouping.item)
> mu <- WB$all.means
> W <- WB$W
> B <- WB$B
> mu
```

```
      logCaK    logCaSi   logCaFe
[1,]  4.20495  -0.7425402 2.770238
```

```
> W
```

```
              logCaK        logCaSi        logCaFe
logCaK  1.688046e-02  2.792714e-05  2.783344e-04
logCaSi 2.792714e-05  6.545540e-05  8.362677e-06
logCaFe 2.783344e-04  8.362677e-06  1.294188e-03
```

```
> B
```

```
             logCaK       logCaSi       logCaFe
logCaK    0.71485025   0.099343866  -0.047824106
logCaSi   0.09934387   0.062724678  -0.007360187
logCaFe  -0.04782411  -0.007360187   0.102438334
```

The Bayes factor can be calculated as the ratio between (3.27) and (3.28) using the function `two.level.mvn.BF` available in the routines file `two_level_functions.r`. This function is part of the supplementary materials available on the website of this book (on `http://link.springer.com/`). First, it is necessary to calculate the sample means $\bar{y}$ and $\bar{x}$ and to determine the sample size n_y and n_x

```
> ybar=as.vector(colMeans(recovered))
> xbar=as.vector(colMeans(control))
> ny=dim(recovered)[1]
> nx=dim(control)[1]
```

(continued)

Example 3.12 (continued)
The Bayes factor can be obtained as

```
> BF=two.level.mvn.BF(W, B, mu, xbar, ybar, nx, ny)
> BF

[1] 157.6265
```

This Bayes factor represents moderately strong support for the proposition according to which the recovered and the control fragments originate from the same source, rather than from different sources. This is expected because the compared measurements refer to the same fragment.

3.4.1.2 Non-normal Distribution for the Between-Source Variability

The two-level random effect model presented in the previous section is based on the assumption of normality of the between-source variability. However, in many practical applications, observations or measurements do not exhibit (enough) regularity for standard parametric models to be used. For example, a multivariate normal distribution for the mean vector $\boldsymbol{\theta}$ may be difficult to justify. It can be replaced by a kernel density estimate, which is sensitive to multimodality and skewness, and which may provide a better representation of the available data.

Starting from a database $\{\mathbf{z}_{ij} = (z_{ij1}, \ldots, z_{ij1}); \ i = 1, \ldots, m \text{ and } j = 1, \ldots, n)\}$, the estimate of the probability density distribution for the between-source variability can be obtained as follows:

$$f(\boldsymbol{\theta} \mid \bar{\mathbf{z}}_1, \ldots, \bar{\mathbf{z}}_m, B, h) = \frac{1}{m} \sum_{i=1}^{m} K(\boldsymbol{\theta} \mid \bar{\mathbf{z}}_i, B, h), \tag{3.35}$$

where the kernel density function $K(\boldsymbol{\theta} \mid \bar{\mathbf{z}}_i, B, h)$ is taken to be a multivariate normal distribution centered at the group mean $\bar{\mathbf{z}}_i$, with covariance matrix $h^2 B$. The smoothing parameter h can be estimated as

$$\hat{h} = \left(\frac{4}{2p+1} \right)^{\frac{1}{p+4}} m^{-1/(p+4)}. \tag{3.36}$$

See also Silverman (1986) and Scott (1992).

We first write a function `hopt` that computes the estimate of the smoothing parameter.

```
> hopt=function(p,m){
+ h=(4/(2*p+1))^(1/(p+4))*m^(-1/(p+4))
+ return(h)}
```

Thus, if the number p of variables is set equal to 4 and the number of sources m is set equal to 30, the smoothing parameter h can be estimated as in (3.36)

```
> p=4
> m=30
> hopt(p,m)

[1] 0.5906593
```

The BF can be obtained as in (3.21), where a multivariate normal distribution is used for the control and the recovered measurements as in (3.22), and a kernel distribution for the between-source variability, as in (3.35). The numerator and the denominator of the BF, $f_1(\mathbf{y}, \mathbf{x} \mid \boldsymbol{\mu}, W, B)$ and $f_2(\mathbf{y}, \mathbf{x} \mid \boldsymbol{\mu}, W, B)$, can be obtained analytically (Aitken and Lucy, 2004). The BF is the ratio between

$$
\mid B \mid^{1/2} m h^p \mid n_y W^{-1}
$$

$$
+ n_x W^{-1} + (h^2 B)^{-1} \mid^{-1/2} \exp\left\{-\frac{1}{2} F_2\right\} \sum_{i=1}^{m} \exp\left\{-\frac{1}{2} F_i\right\} \tag{3.37}
$$

and

$$
\mid n_y W^{-1} + (h^2 B)^{-1} \mid^{-1/2} \sum_{i=1}^{m} \exp\left\{-\frac{1}{2} F_{yi}\right\}
$$

$$
\times \mid n_x W^{-1} + (h^2 B)^{-1} \mid^{-1/2} \sum_{i=1}^{m} \exp\left\{-\frac{1}{2} F_{xi}\right\}, \tag{3.38}
$$

where:

$$
F_i = (\mathbf{w}^* - \bar{\mathbf{z}}_i)' \left\{ \left(n_y W^{-1} + n_x W^{-1}\right)^{-1} + (h^2 B)\right\}^{-1} (\mathbf{w}^* - \bar{\mathbf{z}}_i),
$$

$$
\mathbf{w}^* = \left(n_y W^{-1} + n_x W^{-1}\right)^{-1} \left(n_y W^{-1} \bar{\mathbf{y}} + n_x W^{-1} \bar{\mathbf{x}}\right),
$$

$$
F_{yi} = (\bar{\mathbf{y}} - \bar{\mathbf{z}}_i)' \left(\frac{W}{n_y} + h^2 B\right)^{-1} (\bar{\mathbf{y}} - \bar{\mathbf{z}}_i),
$$

$$
F_{xi} = (\bar{\mathbf{x}} - \bar{\mathbf{z}}_i)' \left(\frac{W}{n_x} + h^2 B\right)^{-1} (\bar{\mathbf{x}} - \bar{\mathbf{z}}_i).
$$

Example 3.13 (Glass Evidence—Continued) Consider the case examined in Example 3.12, and suppose a kernel distribution is used to model the between-source variability (Aitken and Lucy, 2004). Start from the same database, glass-data.txt, covering n replicate measurements of p variables for each of $m = 62$ different sources. The smoothing parameter can be estimated using the function hopt, for $p = 3$.

```
> p=3
> m=62
> h=hopt (p,m)
> h

[1] 0.5119462
```

First, the group means $\bar{z}_i$ must be obtained. They are an output of the function two.level.mv.WB, previously used to estimate the model parameters.

```
> group.means=WB$group.means
```

Here we show only the first six rows of the $(m \times p)$ matrix, where each row represents the means of the measurements $\bar{z}_i = \frac{1}{n} \sum_{i=1}^{n} z_{ij}$.

```
> head(group.means)

     logCaK    logCaSi   logCaFe
2  4.895500  -0.346682  2.445828
3  2.581000  -0.890684  2.922228
4  4.092612  -0.801742  2.761072
5  4.290912  -0.267606  2.665930
6  4.594812  -0.405718  2.674566
7  2.543280  -0.893428  2.898054
```

The Bayes factor can then be calculated as the ratio between (3.37) and (3.38) using the function two.level.mvk.BF contained in the routines file two_level_functions. This function is part of the supplementary materials available on the website of this book (on http://link.springer.com/).

```
> source('two_level_functions.r')
> BF=two.level.mvk.BF(xbar,ybar,nx,ny,W,B, group.
  means, h)
> BF

[1] 151.6001
```

The Bayes factor represents moderately strong support for the proposition according to which the recovered and the control fragments originate from the same source, rather than from different sources.

A detailed comparison and discussion of the performance of these two multivariate random effect models can be found in Aitken and Lucy (2004). An alternative approach to the kernel density estimation is presented by Franco-Pedroso et al. (2016), modeling the between-source distribution by means of a Gaussian mixture model.

Note that a third level of variability could be considered. In fact, one may wish to model separately the variability between replicate measurements from a given item originating from a given source (e.g., replicate measurements from a glass fragment originating from a given window) and the variability between different items originating from a given source (e.g., different glass fragments originating from the same window). This aspect will be tackled in Sect. 3.4.4 where *three-level* models will be introduced.

3.4.1.3 Non-constant Within-Source Variability

The two-level random effect models presented in Sects. 3.4.1.1 and 3.4.1.2 are characterized by the assumption of a constant within-source variability. In other words, it was assumed that every single source has the same intra-variability. While for some type of trace evidence this assumption is acceptable (e.g., for measurements of the elemental composition of glass fragments), a constant within-source variation may be more difficult to justify in other forensic domains. Consider, for example, the case of handwriting on questioned documents where it is largely recognized that intra-variability may vary between writers (Marquis et al., 2006).

Suppose that a handwritten document of unknown source is available for comparative examinations. Handwritten items from a person who is suspected to be the writer are collected and analyzed. Multiple characters are analyzed on the questioned document and on the known writings of the person of interest. The following propositions are defined:

H_1: The person of interest wrote the questioned document.
H_2: An unknown person wrote the questioned document.

The distribution of the vector of means within group (source) θ_i is treated as explained in Sect. 3.4.1.1, i.e., $(\theta_i \mid \mu, B) \sim \mathrm{N}(\mu, B)$. An inverse Wishart distribution is chosen to model the uncertainty about the within-group covariance matrix,

$$(W_i \mid \Omega, v) \sim W^{-1}(\Omega, v), \tag{3.39}$$

where Ω is the scale matrix and v are the degrees of freedom (Bozza et al., 2008). The scale matrix Ω is elicited in a way such that the prior mean of W_i is taken to be equal to the within-group covariance matrix estimated from the available background data as in (3.33), while μ is estimated as in (3.32) and the between-group covariance matrix is estimated as

$$\hat{B} = \frac{1}{m-1} \sum_{i=1}^{m} n(\bar{\mathbf{z}}_i - \bar{\mathbf{z}})(\bar{\mathbf{z}}_i - \bar{\mathbf{z}})'.$$

A two-level multivariate random effect model with an inverse Wishart distribution, modeling the uncertainty about the within-source covariance matrix, has also been proposed by Ommen et al. (2017).

First, consider the numerator of the Bayes factor in (3.21). If proposition H_1 holds, then $\boldsymbol{\theta}_y = \boldsymbol{\theta}_x = \boldsymbol{\theta}$ and $W_y = W_x = W$, and the marginal likelihood is as follows:

$$f(\mathbf{y}, \mathbf{x} \mid H_1) = f_1(\mathbf{y}, \mathbf{x} \mid \boldsymbol{\mu}, B, \Omega, v)$$

$$= \int f(\mathbf{y} \mid \boldsymbol{\theta}, W) f(\mathbf{x} \mid \boldsymbol{\theta}, W) f(\boldsymbol{\theta} \mid \boldsymbol{\mu}, B) f(W \mid \Omega, v) d(\boldsymbol{\theta}, W),$$

$$(3.40)$$

where $f(\boldsymbol{\theta} \mid \boldsymbol{\mu}, B)$ is as in (3.25), and

$$f(W \mid \Omega, v) = \frac{c \mid \Omega \mid^{v-p-1}/2}{\mid W \mid^{v/2}} \exp\left\{-\frac{1}{2}\mathrm{tr}(W^{-1}\Omega)\right\},$$

where c is the normalizing constant (e.g., Press, 2005).

If proposition H_2 holds, then $\boldsymbol{\theta}_y \neq \boldsymbol{\theta}_x$ and $W_y \neq W_x$, and the marginal likelihood takes the following form:

$$f(\mathbf{y}, \mathbf{x} \mid H_2) = f_2(\mathbf{y}, \mathbf{x} \mid \boldsymbol{\mu}, B, \Omega, v) \qquad (3.41)$$

$$= \int f(\mathbf{y} \mid \boldsymbol{\theta}, W) f(\boldsymbol{\theta}, W \mid \boldsymbol{\mu}, B, \Omega, v) d(\boldsymbol{\theta}, W)$$

$$\times \int f(\mathbf{x} \mid \boldsymbol{\theta}, W) f(\boldsymbol{\theta}, W \mid \boldsymbol{\mu}, B, \Omega, v) d(\boldsymbol{\theta}, W).$$

The Bayes factor is the ratio between the marginal likelihoods in (3.40) and (3.41). However, these distributions are not available in closed form as the integrals do not have an analytical solution. Several approaches are available to deal with this problem. Chib (1995) estimates the marginal likelihood $f(\mathbf{y}, \mathbf{x} \mid H_i)$ by a direct application of Bayes theorem, since the marginal likelihood can be seen as the normalizing constant of the posterior density $f(\boldsymbol{\theta}, W \mid \mathbf{y}, \mathbf{x}, H_i)$. The marginal likelihood can therefore be obtained as

$$f(\mathbf{y}, \mathbf{x} \mid H_i) = \frac{f(\mathbf{y}, \mathbf{x} \mid \boldsymbol{\theta}, W) f(\boldsymbol{\theta}, W \mid H_i)}{f(\boldsymbol{\theta}, W \mid \mathbf{y}, \mathbf{x}, H_i)}. \qquad (3.42)$$

While the likelihood function $f(\mathbf{y}, \mathbf{x} \mid \boldsymbol{\theta}, W)$ and the prior density $f(\boldsymbol{\theta}, W \mid H_i)$ can be easily evaluated at any parameter point $(\boldsymbol{\theta}^*, W^*)$, this is not the case for the posterior density $f(\boldsymbol{\theta}, W \mid \mathbf{y}, \mathbf{x}, H_i)$, which is not known in closed form. A Gibbs sampling algorithm (Sect. 1.8) can be applied to the set of the complete conditional densities $f(\boldsymbol{\theta} \mid W, \mathbf{y}, \mathbf{x}, H_i)$ and $f(W \mid \boldsymbol{\theta}, \mathbf{y}, \mathbf{x}, H_i)$, and the posterior density $f(\boldsymbol{\theta}, W \mid \mathbf{y}, \mathbf{x}, H_i)$ can be approximated from the output of the Gibbs sampling algorithm as $\hat{f}(\boldsymbol{\theta}, W \mid \mathbf{y}, \mathbf{x}, H_i)$ (Chib, 1995; Bozza et al., 2008; Aitken et al., 2021).

The marginal likelihood in (3.42) can be estimated at a given parameter point $(\boldsymbol{\theta}^*, W^*)$ as

$$\hat{f}(\mathbf{y}, \mathbf{x} \mid H_i) = \frac{f(\mathbf{y}, \mathbf{x} \mid \boldsymbol{\theta}^*, W^*) f(\boldsymbol{\theta}^*, W^* \mid H_i)}{f(\boldsymbol{\theta}^*, W^* \mid \mathbf{y}, \mathbf{x}, H_i)}.$$

The Bayes factor is then calculated as

$$\mathrm{BF} = \frac{\hat{f}(\mathbf{y}, \mathbf{x} \mid H_1)}{\hat{f}(\mathbf{y}, \mathbf{x} \mid H_2)}. \tag{3.43}$$

As mentioned in Sect. 1.8, many other approaches are available, and their efficiency should be studied and compared.

Example 3.14 (Handwriting Evidence)   Consider a hypothetical case involving a handwritten document. Handwritten items from a person of interest are available for comparative examinations. The propositions of interest are therefore:

H_1 :   The person of interest wrote the questioned document.
H_2 :   An unknown person wrote the questioned document.

Suppose that $n_1 = 8$ characters of type a are collected from the questioned document and that $n_2 = 8$ characters of the same type are extracted from a document originating from the person of interest, taken for comparative purposes. The contour shape of loops of handwritten characters can be described using a methodology based on Fourier analysis (Marquis et al., 2005, 2006). In brief, the contour shape of each handwritten character loop can be described by means of a set of variables representing the surface and a set of harmonics. Each harmonic corresponds to a specific contribution to the shape and is defined by an amplitude and a phase, the Fourier descriptors.

Consider the database named `handwriting.txt` available on the book's website. It contains data on $p = 9$ variables (i.e., the surface, the amplitude and the phase of the first four harmonics), measured on several characters of type a collected from $m = 20$ writers. The variables of interest are displayed in columns 2 to 10. Column 1 contains the item (writer) identifier

(continued)

Example 3.14 (continued)

```
> population=read.table('handwriting.txt',
+ header=TRUE)
> names(population)=c('writer','A0','A1','B1','A2',
+'B2','A3','B3','A4','B4')
> variables=2:10
> grouping.item=1
```

In the current example, measurements **y** on the questioned document and measurements **x** on the control document were randomly selected from the available measurements on characters collected from a given writer (i.e., writer no. 1). Starting from a total number of, say, n available characters, $2 \times n_1$ characters have been selected: the first n_1 characters serve as recovered data, while the remaining serve as control data

```
> item=1
> base=population[which(population[,grouping.item]
+ ==item),]
> nr=dim(base)[1]
> n1=8
> recovered=as.matrix(base[1:n1,variables])
> control=as.matrix(base[(n1+1):(2*n1),variables])
```

Data concerning measurements from the selected writer were then excluded from the database

```
> pop.back=population[-which(population[,grouping.
+ item]==item),]
```

The database pop.back will serve as background data and can be used to estimate the model parameters as in (Bozza et al., 2008) using the function two.level.mv.WB available in the file two_level_functions.r.

```
> source('two_level_functions.r')
> WB = two.level.mv.WB(pop.back,variables,
+ grouping.item,nc=TRUE)
> mu = t(WB$all.means)
> W = WB$W
> B = WB$B
```

The number of degrees of freedom ν of the inverse Wishart distribution is chosen so as to reduce the variability of this distribution, centered at the within-source covariance matrix estimated as in (3.33).

```
> p=9
> nu=40
> Omega=W*(nu-2*p-2)
```

(continued)

Example 3.14 (continued)
The Gibbs sampling algorithm is run over 10000 iterations with a burn-in of 1000.

```
> n.iter=10000
> burn.in=1000
```

The Bayes factor in (3.43) can then be calculated using the function two.level.mvniw.BF that is part of the supplementary materials. Note also that this routine requires other routines that are available in the packages MCMCpack (Martin et al., 2021) and mvtnorm (Genz et al., 2020).

```
> BF=two.level.mvniw.BF(recovered,control,Omega,B,mu,
+ nu,p, n.iter,burn.in)
> BF
```

```
[1] 5543330
```

The Bayes factor represents extremely strong support for the proposition according to which the questioned and the recovered handwritten materials originate from the same source, rather than from different sources. A fully documented open-source package (Gaborini, 2019) has been developed by Gaborini (2021).

Note that it is important to critically examine large BF values, such as the one obtained above. For a discussion about extreme values, see Aitken et al. (2021), Hopwood et al. (2012), and Kaye (2009). Moreover, as underlined in Sect. 1.11, the marginal likelihood is highly sensitive to the prior assessments and so is the BF. In particular, while the overall mean vector, the within- and the between-source covariance matrices are estimated from the available background data, the number of degrees of freedom of the inverse Wishart distribution are chosen so as to reduce the dispersion of the prior. A sensitivity analysis may be performed to assess the sensitivity of the BF to different choices of the degrees of freedom v in (3.39).

The BF may also be sensitive to the MCMC approximation. Figure 3.5 provides an illustration of BF variability. Results are based on 50 realizations of the BF approximation in (3.43).

```
> ns=50
> BFs=matrix(0,nrow=ns,ncol=1)
> for(i in 1:ns){
+ BFs[i]=two.level.mvniw.BF(recovered,control,Omega,B,
+ mu,nu,p,n.iter,burn.in)}
> hist(log(BF),freq=F,main='',xlab='log(BF)')
```

Fig. 3.5 Histogram of 50
realizations of the BF
approximation in (3.43)

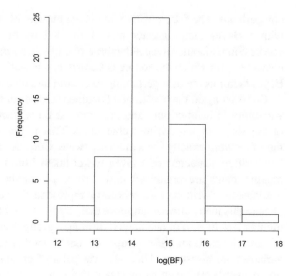

The models discussed here rely on the assumption of independence between sources, focusing on the inherent variability of features. In the case of questioned documents (Sect. 3.4.1.3), this amounts to assume that handwritten material has been produced without any intention of reproducing someone else's writing style. The possibility of forgery and/or disguise breaks the independence assumption made at denominator. Section 3.4.3 will address this complication.

3.4.2   Assessment of Method Performance

The results of the procedures described in the previous sections may be sensitive to changes in the features of recovered and control materials, the available background information, as well as to choices made during probabilistic modeling and prior elicitation. A sensitivity analysis may be conducted in order to gain a better understanding of the properties of the chosen method. It is fundamental to gain an understanding of how well a method performs: if the recovered and control data originate from the same source, the BF is expected to be greater than 1. Vice versa, if the compared items come from different sources, a BF smaller than 1 is expected.

Several methods exist for the assessment of the performance of the methods for evidence evaluation. Commonly encountered measures in this context are rates of false negatives (i.e., cases in which the Bayes factor is smaller than 1, supporting hypothesis H_2, when hypothesis H_1 holds) and false positives (i.e., cases in which the Bayes factor is greater than 1, supporting hypothesis H_1, when hypothesis H_2 holds). The rate of false negatives is the number of same-source comparisons with a Bayes factor smaller than 1 divided by the total number of same-source comparisons. The false positive rate is the number of different-source comparisons with a Bayes factor greater than 1 divided by the total number of different-source comparisons. Given a database of cases (e.g., measurements on handwriting characters) for which the source is known, it is possible to study the behavior of the Bayes factor as the data pertaining to control and recovered items change.

Consider again the questioned document case discussed in Sect. 3.4.1.3. There is variability in handwriting, and the reported Bayes factor is sensitive to variability of the shape of handwritten characters. This is not surprising as no one writes the same word exactly the same way twice. Consider measurements of features of handwritten characters of a given writer taken from the available database. These measurements are organized into a $(n \times p)$ matrix, where n is the number of available handwritten characters and p represents the number of features (variables). Denote this matrix base. Suppose that, among the n characters, we select a certain number $2 \times n_1 < n$ of characters, forming a group. Repeating this a certain number of times leads to multiple groups. On each member (character) within a group, p variables are measured. Then we take pairs of groups (i.e., measurements on the group members), taken to represent recovered and control data. Then, the Bayes factor is calculated for each couple. Here, each couple represents a same-source comparison.

Example 3.15 (Two-Level Model for Handwriting—Assessment of Model Performance) Recall Example 3.14 where a total number of 16 characters have been randomly selected from the available characters collected from a given writer (writer no. 1), extracted from the database handwriting.txt. A Bayes factor equal to 5543330 was obtained. If different sets of characters are extracted, the Bayes factor will be influenced (also) by the within-writer variability.

Suppose now that, for the same writer, $ns = 50$ distinct groups of characters (each of size 16) are drawn and split into groups of size 8 to act as questioned and control data. The Bayes factor is calculated for each of the 50 groups. Clearly, since the sampled measurements originate from the same writer, we expect Bayes factors greater than 1.

```
> ns=50
> n=dim(base)[1]
> n1=8
> BFs=matrix(0,nrow=ns,ncol=1)
> for (i in 1:ns){
+         ind=sample(1:n,2*n1,replace=F)
+         recovered=as.matrix(base[ind[1:n1],
+         variables])control=as.matrix(base
+         [ind[(n1+1):length(ind)],variables])
```

(continued)

Example 3.15 (continued)
```
+             BFs[i]=two.level.mvniw.BF(recovered,
+             control,Omega,
+             B,mu,nu,p,n.iter,burn.in)
+             }
```

Figure 3.6 shows a histogram of the results for the $ns = 50$ groups of sampled characters. No false negatives have been observed. The range of the BF values obtained is given here below

```
> range(BFs)

[1] 1.709027e+02 1.438262e+29
```

There is also variability between writers, as no two writers write exactly alike. Consider now measurements of features of handwritten characters from a different writer, say writer no. 6, drawn from the same database. These measurements are stored in a matrix denoted base2.

```
> item2=6
> base2=population[which(population[,grouping.item]==
+ item2),]
> n2=dim(base2)[1]
```

We first estimate the population parameters from the background population where both selected writers have been eliminated.

```
> pop.back=population[-which(population[,grouping
+ .item]==item|population[,grouping.item]==item2),]
> WB = two.level.mv.WB(pop.back,variables,
+ grouping.item,nc=TRUE)
> mu = t(WB$all.means)
> W = WB$W
> B = WB$B
> Omega=W*(nu-2*p-2)
```

Next, for each of the two writers, take 50 groups of characters (from base and base2). Each group contains 8 members, on each of which p features are measured. Then, take a group from each writer and form a so-called known different-source pair, and do this multiple times. These draws are taken to represent recovered and control data. Then, the Bayes factor is calculated for each couple.

```
> ns=50
> n=dim(base)[1]
> nc=dim(base2)[1]
> n1=8
```

(continued)

Fig. 3.6 Histogram of
log(BF) values for 50 groups,
each containing 8 handwritten
characters, sampled from a
given writer to act as
questioned and control
datasets

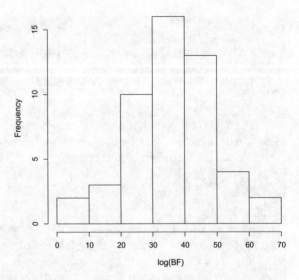

Example 3.15 (continued)

```
> BFs2=matrix(0,nrow=ns,ncol=1)
> for (i in 1:ns){
+          val.r=sample(1:n,n1)
+          recovered=as.matrix(base[val.r,variables])
+          val.c=sample(1:nc,n1)
+          control=as.matrix(base2[val.c,variables])
+          BFs[i]=two.level.mvniw.BF(recovered,
+          control,Omega,B,
+          mu,nu,p,n.iter,burn.in)
+          }
```

Figure 3.7 shows a histogram of the results. No false positives have been
observed. The range of the BF values obtained is

```
> range(BFs)
```

```
[1]  2.733273e-10 7.034354e-02
```

The variability of BF values for different samples is not surprising because of
handwriting variability. However, this should not be understood as there being a
Bayes factor distribution. See, e.g., Morrison (2016), Ommen et al. (2016), and
Taroni et al. (2016) for a discussion of issues relating to the reporting of the precision
of forensic likelihood ratios.

Over the past decade, several other approaches have been proposed in forensic
statistics literature for evaluating the performance of statistical procedures, based

Fig. 3.7 Histogram of log(BF) values obtained for 50 groups, each containing 8 handwritten characters, sampled from the same couple of writers to act as questioned and control datasets

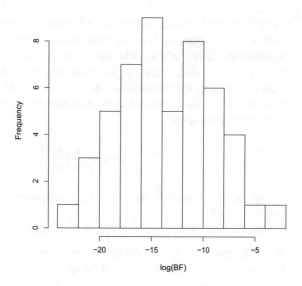

on a likelihood ratio or a Bayes factor. These methods provide a rigorous approach to assessing and comparing the performance of evaluative methods prior to using them in casework and forensic reporting. See, in particular, Ramos and Gonzalez-Rodriguez (2013) and Ramos et al. (2021) for a methodology to measure calibration of a set of likelihood ratio values and the concept of Empirical Cross-Entropy for representing performance, illustrated using examples from forensic speech analysis. These concepts are also discussed by Meuwly et al. (2017) who present a guideline for the validation of evaluative methods considering source level propositions. Zadora et al. (2014) present performance assessment for physicochemical data in the context of trace evidence (e.g., glass). For a recent review, see also Chapter 8 of Aitken et al. (2021).

3.4.3 On the Assumption of Independence Under H_2

The models presented in Sect. 3.4.1 are based on the assumption of independence between the questioned and known materials under hypothesis H_2. This may be reasonable for certain types of evidence and cases, but less for others. In fact, while a physical feature (e.g., the elementary composition of glass fragments) requires external constraint to be altered, a behavioral or biometric feature such as signature can be modified intentionally.

Consider handwriting as an example. When evaluating results of comparative handwriting examination, the case circumstances may be such that there is no issue of handwriting features being disguised or the result of an attempt to imitate the handwriting of another person. The approach suggested in Sect. 3.4.1.3 may thus be applicable. In turn, in case of alleged forgery of signatures, the (unknown) writer specifically intends to reproduce features of a target signature. The allegation, then, is that a signature is either simulated or disguised, rather than presenting a correspondence or similarity with a genuine signature by mere chance alone (Linden et al., 2021). In such cases, the Bayes factors previously developed in Sect. 3.4.1 cannot be used to approach the question of interest here because the assumption of independence between sources at the denominator cannot be maintained. It follows that one must compute

$$\mathrm{BF} = \frac{f(\mathbf{y} \mid \mathbf{x}, H_1)}{f(\mathbf{y} \mid \mathbf{x}, H_2)}, \tag{3.44}$$

as $f(\mathbf{y} \mid \mathbf{x}, H_2)$, following the above argument, does not simplify to $f(\mathbf{y} \mid H_2)$ (see also Sect. 1.5.1).

Consider the following competing propositions:

H_1 : The person of interest (POI) produced the questioned signature.
H_2 : An unknown person produced the questioned signature, trying to simulate the POI's signature.

If proposition H_2 is true, the forensic document examiner has to deal with a signature written by someone who has knowledge of the POI's signature.

Consider the two-level model in Sect. 3.4.1.3 where the distribution of the measurements on the recovered and control data is taken to be Normal, with vector means $\boldsymbol{\theta}_y$ and $\boldsymbol{\theta}_x$, and covariance matrices W_y and W_x

$$(Y \mid \boldsymbol{\theta}_y, W_y) \sim \mathrm{N}(\boldsymbol{\theta}_y, W_y) \qquad ; \qquad (X \mid \boldsymbol{\theta}_x, W_x) \sim \mathrm{N}(\boldsymbol{\theta}_x, W_x). \tag{3.45}$$

The probability densities at the numerator and denominator of the BF in (3.44) can be obtained as

$$f(\mathbf{y}, \mathbf{x} \mid H_i) = f_i(\mathbf{y}, \mathbf{x} \mid \boldsymbol{\mu}_i, B_i, \Omega_i, \nu_i)$$

$$= \int f(\mathbf{y} \mid \boldsymbol{\theta}, W) f(\boldsymbol{\theta}, W \mid \mathbf{x}, \boldsymbol{\mu}_i, B_i, \Omega_i, \nu_i), \tag{3.46}$$

where $(\boldsymbol{\mu}_i, B_i)$ and (Ω_i, ν_i) are the hyperparameters of the prior distributions under the competing propositions (i.e., a normal prior and an inverse Wishart prior distribution). The Bayes factor can thus be calculated as

$$\mathrm{BF} = \frac{f_1(\mathbf{y}, \mathbf{x} \mid \boldsymbol{\mu}_1, B_1, \Omega_1, \nu_1)}{f_2(\mathbf{y}, \mathbf{x} \mid \boldsymbol{\mu}_2, B_2, \Omega_2, \nu_2)}. \tag{3.47}$$

Two different background databases are needed to inform model parameters under the competing propositions: a database of genuine signatures ($\mathbf{z}_{ij}$) and a database of imitated signatures ($\mathbf{s}_{ij}$). Someone who imitates a signature needs to work outside their writing habits and movement patterns. Thus, simulated signatures do not reflect the same movements and writing features as genuine signatures. Model parameter μ_i can be estimated as in (3.32), and B_i as explained in Sect. 3.4.1.3. The scale matrix Ω_i can be chosen so as to center the prior distribution at the within-group covariance matrix W_i that can be estimated as in (3.33).

The probability densities in (3.46) are not available in closed form but can be estimated from the output of a MCMC algorithm following, for example, the ideas described in Sect. 3.4.1.3. A Gibbs sampling algorithm is implemented here. The routine is different from that developed in Sect. 3.4.1.3 because it calculates the BF in (3.47). In this formula, no assumption of independence is made at the denominator, and two different databases are used.

Example 3.16 (Digitally Captured Signatures)   Consider a case involving a questioned signature on a contract signed on a digital tablet. The person of interest denies having signed the contract. Among the multiple features that are captured by the digital tablet, the average speed and writing time are considered here. See Linden et al. (2021) for a detailed description of the experimental conditions. Measurements on the questioned signature are $y = (4639, 380.42)$, while measurements on the control signature are $x = (4460, 323.4787)$. Note that the first value is the average speed and the second is the writing time.

```
> quest=c(4639,380.42)
> ref=c(4460,323.4787)
```

Model parameters under hypothesis H_1 (i.e., the mean vector μ_1, the within-group covariance matrix W_1, and the between-group covariance matrix B_1) are estimated from an available database of genuine signatures (z_{ij}) and are given here below.

```
> mug=matrix(c(2754.767,511.284),ncol=1)
> Wg=matrix(c(95755.861,-4214.939,-4214.939,
+ 2857.975),byrow=T,nrow=2)
> Bg=matrix(c(3377136,30548.24,30548.24,20335.10),
+ byrow=T,nrow=2)
```

The trace matrix of the inverse Wishart distribution is then obtained as

```
> p=2
> nu=10
> Omegag=Wg*(nu-2*p-2)
```

In the same way, model parameters under hypothesis H_2 are estimated from an available database of simulated signatures (s_{ij}) and are given here below.

(continued)

Example 3.16 (continued)
```
> mus=matrix(c(14824.3,145.0719),ncol=1)
> Ws=matrix(c(14798844,-42412.0995,-42412.0995,
+ 940.0561), byrow=T,nrow=2)
> Bs=matrix(c(37657528.8,-157142.437,-157142.437,
+ 3691.482), byrow=T,nrow=2)
> Omegas=Ws*(nu-2*p-2)
```

A Gibbs sampling algorithm is run over 10000 iterations, with a burn-in of 1000.

```
> n.iter=10000
> burn.in=1000
```

The Bayes factor in (3.44) can then be calculated using the function two.level.mvniw2.BF (see supplementary materials).

```
> source('two_level_functions.r')
> BF=two.level.mvniw2.BF(quest,ref,Wg,Bg,mug,Ws,Bs,
+ mus,nu,p,n.iter,burn.in)
> BF
```

```
[1] 40846.87
```

The BF represents very strong support for the proposition according to which the questioned signature originates from the person of interest rather than from an unknown person who attempted to imitate the target signature.

3.4.4 Three-Level Models

So far, two-level models have been considered, taking into account the within-source and the between-source variability. However, it is not uncommon to encounter situations in which the hierarchical ordering shows an additional level of variability, e.g., in relation to measurement error.

Denote again by p the number of variables observed on items of a given evidential type. Suppose that continuous measurements of these variables are available on a random sample from m sources with s items for each source and n replicate measurements on each of the $N = ms$ items. The background data can be denoted by $\mathbf{z}_{ikj} = (z_{ikj1}, \ldots, z_{ikjp})'$, where $i = 1, \ldots, m$ denotes the number of sources (e.g., windows, writers), $k = 1, \ldots, s$ denotes the number of items for each source (e.g., glass fragments, handwritten characters), and $j = 1, \ldots, n$ denotes the number of replicate measurements for each item.

A Bayesian statistical model for the evaluation of evidence for three-level normally distributed multivariate data was proposed by Aitken et al. (2006), focusing on the elemental composition of glass fragments. Denote the mean vector within item k in group i as $\boldsymbol{\theta}_{ik}$ and the covariance matrix of replicate measurements as W. For the variability of replicate measurements, the distribution of $\mathbf{Z}_{ikj}$ is taken to be normal, $\mathbf{Z}_{ikj} \sim \mathrm{N}(\boldsymbol{\theta}_{ik}, W)$.

Denote by $\boldsymbol{\mu}_i$ the mean vector within group i and by V the within-group covariance matrix. The distribution of $\boldsymbol{\theta}_{ik}$ for the within-group variability is taken to be normal, $\boldsymbol{\theta}_{ik} \sim \mathrm{N}(\boldsymbol{\mu}_i, B)$.

Denote by $\boldsymbol{\phi}$ the mean vector between groups. Let U denote the between-group covariance matrix. For the between-group variability, the distribution of the $\boldsymbol{\mu}_i$ is taken to be normal, $\boldsymbol{\mu}_i \sim \mathrm{N}(\boldsymbol{\phi}, V)$.

Consider the case described in Sect. 3.4.1, where measurements are available on n_y items from an unknown origin as well as measurements on n_x items from a known origin. These two groups of items may or may not come from the same source. Competing propositions may be formulated as follows:

H_1 : The recovered and the control items originate from the same source.

H_2 : The recovered and the control items originate from different sources.

There are n_1 replicate measurements available on each of the recovered n_y items. Denote the measurement vector by $\mathbf{y}$, where the vector components are denoted by $\mathbf{y}_{kj}$ (for $k = 1, \ldots, n_y$ and $j = 1, \ldots, n_1$) and $\mathbf{y}_{kj} = (y_{kj1}, \ldots, y_{kjp})'$. For each of the n_x control items, n_2 replicate measurements are available. Denote the measurement vector by $\mathbf{x}$, where the vector components are denoted $(\mathbf{x}_{kj}, k = 1, \ldots, n_x$ and $j = 1, \ldots, n_2)$ and $\mathbf{x}_{kj} = (x_{kj1}, \ldots, x_{kjp})'$.

The Bayes factor is the ratio of two probability densities of the form $f(\mathbf{y}, \mathbf{x} \mid H_i) = f_i(\mathbf{y}, \mathbf{x} \mid \boldsymbol{\phi}, W, B, V)$, $i = 1, 2$. The probability density in the numerator is given by

$$f_1(\mathbf{y}, \mathbf{x} \mid \boldsymbol{\phi}, W, B, V)$$
$$= \int \int f(\mathbf{y} \mid \boldsymbol{\theta}, W) f(\mathbf{x} \mid \boldsymbol{\theta}, W) f(\boldsymbol{\theta} \mid \boldsymbol{\mu}, B) f(\boldsymbol{\mu} \mid \boldsymbol{\phi}, V) d\boldsymbol{\mu} d\boldsymbol{\theta}, \quad (3.48)$$

where all probability densities are multivariate normal.

In the denominator, the probability density is given by

$$f_2(\mathbf{y}, \mathbf{x} \mid \boldsymbol{\phi}, W, B.V) = \int \int f(\mathbf{y} \mid \boldsymbol{\theta}, W) f(\boldsymbol{\theta} \mid \boldsymbol{\mu}, B) f(\boldsymbol{\mu} \mid \boldsymbol{\phi}, V) d\boldsymbol{\mu} d\boldsymbol{\theta}$$
$$\times \int \int f(\mathbf{x} \mid \boldsymbol{\theta}, W) f(\boldsymbol{\theta} \mid \boldsymbol{\mu}, B) f(\boldsymbol{\mu} \mid \boldsymbol{\phi}, V) d\boldsymbol{\mu} d\boldsymbol{\theta}, \quad (3.49)$$

where all probability densities are multivariate normal.

As shown by Aitken et al. (2006), the value of the evidence is the ratio of

$$
\mid B + V \mid^{1/2} \mid [(n_y n_1 + n_x n_2) W^{-1}
$$

$$
+ (B + V)^{-1}] \mid^{-1/2} \exp \left\{ -\frac{1}{2}(F_1 + F_2) \right\} \tag{3.50}
$$

to

$$
\mid (n_y n_1 W^{-1} + (B + V)^{-1}) \mid^{-1/2} \mid n_x n_2 W^{-1} + (B + V)^{-1} \mid^{-1/2}
$$

$$
\times \exp \left\{ -\frac{1}{2}(F_3 + F_4) \right\}, \tag{3.51}
$$

where:

$$
F_1 = (\bar{\mathbf{y}} - \bar{\mathbf{x}})' \left(\frac{n_y n_1 n_x n_2 W^{-1}}{n_y n_1 + n_x n_2} \right) (\bar{\mathbf{y}} - \bar{\mathbf{x}}),
$$

$$
F_2 = (\bar{\mathbf{w}} - \boldsymbol{\phi})' \left((n_y n_1 + n_x n_2)^{-1} W + B + V \right)^{-1} (\bar{\mathbf{w}} - \boldsymbol{\phi}),
$$

$$
F_3 = (\bar{\mathbf{y}} - \boldsymbol{\phi})' \left[(n_y n_1)^{-1} W + B + V \right]^{-1} (\bar{\mathbf{y}} - \boldsymbol{\phi}),
$$

$$
F_4 = (\bar{\mathbf{x}} - \boldsymbol{\phi})' \left[(n_x n_2)^{-1} W + B + V \right]^{-1} (\bar{\mathbf{x}} - \boldsymbol{\phi}),
$$

and $\bar{\mathbf{w}} = \frac{n_y n_1 \bar{\mathbf{y}} + n_x n_2 \bar{\mathbf{x}}}{n_y n_1 + n_x n_2}$.

The overall mean $\boldsymbol{\phi}$, the measurement error covariance matrix W, the within-group covariance matrix B, and the between-group covariance matrix V can be estimated using the available background data:

$$
\hat{\boldsymbol{\phi}} = \frac{1}{m} \frac{1}{s} \frac{1}{n} \sum_{i=1}^{m} \sum_{k=1}^{s} \sum_{j=1}^{n} \mathbf{z}_{ikj}, \tag{3.52}
$$

$$
\hat{W} = \frac{1}{ms(n-1)} \sum_{i=1}^{m} \sum_{k=1}^{s} \sum_{j=1}^{n} (\mathbf{z}_{ikj} - \bar{\mathbf{z}}_{ik.})(\mathbf{z}_{ikj} - \bar{\mathbf{z}}_{ik.})', \tag{3.53}
$$

$$
\hat{B} = \frac{1}{m(s-1)} \sum_{i=1}^{m} \sum_{k=1}^{s} (\bar{\mathbf{z}}_{ik.} - \bar{\mathbf{z}}_{i..})(\bar{\mathbf{z}}_{ik.} - \bar{\mathbf{z}}_{i..})' - \frac{\hat{W}}{n}, \tag{3.54}
$$

$$
\hat{V} = \frac{1}{m-1} \sum_{i=1}^{m} (\bar{\mathbf{z}}_{i..} - \bar{\mathbf{z}}_{...})(\bar{\mathbf{z}}_{i..} - \bar{\mathbf{z}}_{...})' - \frac{\hat{B}}{s} - \frac{\hat{W}}{sn}, \tag{3.55}
$$

where $\bar{\mathbf{z}}_{ik.} = \frac{1}{n} \sum_{j=1}^{n} \mathbf{z}_{ikj}, \bar{\mathbf{z}}_{i..} = \frac{1}{s} \sum_{k=1}^{s} \mathbf{z}_{ik.}$ and $\bar{\mathbf{z}}_{i...} = \frac{1}{m} \sum_{i=1}^{m} \bar{\mathbf{z}}_{i..}$

Example 3.17 (Glass Evidence—Continued) Consider again the case described in Example 3.12 where two glass fragments are recovered on the jacket of an individual who is suspected to be involved in a crime. Two glass fragments are collected at the crime scene for comparative purposes. The competing propositions are:

(continued)

Example 3.17 (continued)

H_1 : The recovered and known glass fragments originate from the same source (e.g., a broken window).

H_2 : The recovered and known glass fragments originate from different sources.

A database named `glass-database.txt` is available as part of the supplementary material of Zadora et al. (2014). It contains measurements of the elemental concentration of glass fragments from several windows ($m = 200$). For each source, there are $s = 12$ fragments with $n = 3$ replicate measurements. For each fragment, five variables are considered: the logarithmic transformation of the ratios $Na/O, Mg/O, Al/O, Si/O, Ca/O$. The variables of interest are displayed in columns $3, 4, 5, 6$, and 8, while the object (window) identifier is in column 1. The fragment identifier is in column 2.

```
> population=read.table('glass-database.txt',
+ header=T)
> variables=c(3,4,5,6,8)
> grouping.item=1
> grouping.fragment=2
```

Three replicate measurements are available for each fragment. Using the notation introduced above

```
> ny=2
> nx=2
> n1=3
> n2=3
```

Measurements for the recovered fragments, **y**, and measurements for the control fragments, **x**, were selected from the available data for the first and second group (window) and the first two items (fragments) from these windows. Therefore, a BF smaller than 1 is expected.

```
> recovered.item=1
> control.item=2
> base_c=population[which(population[,grouping.item]
+ ==control.item),]
> base_r=population[which(population[,grouping.item]
+ ==recovered.item),]
> recovered=base_r[which(base_r[,grouping.fragment]
+ ==1|base_r[,grouping.fragment]==2),
+ c(2,variables)]
> recovered
```

(continued)

Example 3.17 (continued)

```
  fragment   logNaO   logMgO   logAlO   logSiO
1        1  -0.6603  -1.4683  -1.4683  -0.1463
2        1  -0.6658  -1.4705  -1.4814  -0.1429
3        1  -0.6560  -1.4523  -1.4789  -0.1477
4        2  -0.6309  -1.4707  -1.5121  -0.1823
5        2  -0.6332  -1.4516  -1.4996  -0.1792
6        2  -0.6315  -1.4641  -1.4883  -0.1710
   logCaO
1 -1.1096
2 -1.1115
3 -1.1118
4 -1.1306
5 -1.1332
6 -1.1291

> control=base_c[which(base_c[,grouping.fragment]==1|
+ base_c[,grouping.fragment]==2),c(2,variables)]
> control

   fragment   logNaO   logMgO   logAlO   logSiO
13       1  -0.6231  -1.3641  -1.6540  -0.0964
14       1  -0.6122  -1.3589  -1.6622  -0.0886
15       1  -0.6108  -1.3742  -1.6935  -0.1205
16       2  -0.6135  -1.3686  -1.7202  -0.1381
17       2  -0.6205  -1.3844  -1.6831  -0.1273
18       2  -0.6204  -1.3692  -1.7269  -0.1199
    logCaO
13 -0.9993
14 -0.9836
15 -1.0524
16 -1.0830
17 -1.0721
18 -1.0392
```

Next, the means of measurements $\bar{\mathbf{y}}$, $\bar{\mathbf{x}}$, and $\bar{\mathbf{w}}$ are obtained.

```
> bary=colMeans(recovered[,-1])
> barx=colMeans(control[,-1])
> barw=colMeans(rbind(recovered,control)[,-1])
```

(continued)

Example 3.17 (continued)

Data concerning measurements from the first two windows were then excluded from the database

```
> pop.back <- population[-which(population[,
+ grouping.item]==1|population[,grouping.item]==2),]
```

The database named pop.back will serve as background data. It can be used to estimate the model parameters ϕ, W, B, and V as in (3.52), (3.53), (3.54) and (3.55) by means of the function three.level.mv.WBV contained in the routines file three_level_functions.r. This file is part of the supplementary materials available on the book's website and can be run in the R console with the command

```
> source('three_level_functions.r')
```

The overall mean, the measurement error covariance matrix, the within-source covariance matrix, and the between-source covariance matrix can be estimated as follows:

```
> WBV=three.level.mv.WBV(pop.back,variables,
+ grouping.item,grouping.fragment)
> psi=WBV$overall.means
> W=WBV$W
> B=WBV$B
> V=WBV$V
```

The Bayes factor can be calculated as the ratio between (3.50) and (3.51) using the function three.level.mvn.BF available in the routines file three_level_functions.r. This function is part of the supplementary materials available on the book's website.

```
> BF=three.level.mvn.BF(bary,barx,barw,ny,nx,n1,n2,
+ psi,W,B,V)
> BF
```

```
[1] 0.000083299
```

The Bayes factor represents extremely strong support for the proposition according to which the recovered and the control fragments originate from different sources, rather than from the same source.

Note that the above development does not take into account the topic of variable selection. See Aitken et al. (2006) for a proposal for dimensionality reduction based on a probabilistic structure, determined by a graphical model obtained from a scaled inverse covariance matrix.

3.5   Summary of R Functions

The R functions outlined below have been used in this chapter.

Functions Available in the Base Package

`colMeans`: Forms column means for numeric arrays (or data frames)

`d <name of distribution>`, `p <name of distribution>` (e.g., `dpois`, `pnorm`): Calculate the density and the cumulative probability for many parametric distributions.

More details can be found in the Help menu, `help.start()`.

Functions Available in Other Packages

`dinvgamma` in package `extraDistr`: calculates the density of an inverse gamma distribution.

`dstp` in package `LaplacesDemon`: calculates the density of a non-central Student t distribution.

Functions Developed in the Chapter

`hopt`: Calculates the estimates $\hat{h}$ of the smoothing parameter h.
Usage: `hopt(p,m)`.
Arguments: p, the number of variables: m, the number of sources.
Output: A scalar value.

`poisg`: Computes the density of a Poisson–gamma distribution $Pg(\alpha, \beta, 1)$ at x.
Usage: `poisg(a,b,x)`.
Arguments: a, the shape parameter α; b, the rate parameter β; x, a scalar value x.
Output: A scalar value.

`post_distr`: Computes the posterior distribution $N(\mu_x, \tau_x^2)$ of a normal mean θ, with $X \sim N(\theta, \sigma^2)$ and $\theta \sim N(\mu, \tau^2)$.
Usage: `post_distr(sigma,n,barx,pm,pv)`.
Arguments: `sigma`, the variance σ^2 of the observations; n, the number of observations; `barx`, the sample mean $\bar{x}$ of the observations; pm, the mean μ of the prior distribution $N(\mu, \tau^2)$; pv, the variance τ^2 of the prior distribution $N(\mu, \tau^2)$.
Output: A vector of values, the first is the posterior mean μ_x, the second is the posterior variance τ_x^2.

`two.level.mv.WB`: Computes the estimate of the overall mean $\boldsymbol{\mu}$, the group means $\bar{\mathbf{z}}_i$, the within-group covariance matrix W, and the between-group covariance matrix B for the two-level model in Sect. 3.4.1.
Usage: `two.level.mv.WB(population, variables, grouping.variable,nc=FALSE)`.
Arguments: `population`, a data frame with N rows and k columns for measurements on m sources with n items for each source; `variables`, a vector con-

taining the column indices of the variables to be used; `grouping.variable`, a scalar specifying the variable that is to be used as the grouping factor. By default (nc = FALSE), the between-group covariance matrix is estimated as in Sect. 3.4.1.1. If nc = TRUE, the between-group covariance matrix is estimated as in Sect. 3.4.1.3.

Output: The group means $\bar{z}_i$, the estimated overall mean $\hat{\mu}$, the estimated within-group covariance matrix $\hat{W}$, the estimated between-group covariance matrix $\hat{B}$.

`two.level.mvn.BF`: Computes the BF for a two-level random effect model where both the within-source variability and the between-source variability are normally distributed, and the within-source covariance matrix is constant between sources.

Usage: `two.level.mvn.BF(W,B,mu,xbar,ybar,nx,ny)`.

Arguments: W, the within-source covariance matrix; B, the between-source covariance matrix; mu, the mean vector between sources; xbar, the vector of means for the control item; ybar, the vector of means for the recovered item; nx, the number of measurements for the control material; ny, the number of measurements for the recovered material.

Output: A scalar value.

`two.level.mvk.BF`: Computes the BF for a two-level random effect model where the within-source variability is normally distributed, the normal distribution for the between-source variability is replaced by a kernel density distribution, and the within-source covariance matrix is constant between sources.

Usage: `two.level.mvk.BF(xbar,ybar,nx,ny,W,B,group.means,h)`.

Arguments: xbar, the vector of means for the control item; ybar, the vector of means for the recovered item; nx, the number of measurements for the control material; ny, the number of measurements for the recovered material; W, the within-source covariance matrix; B, the between-source covariance matrix; group.means, a ($m \times p$) matrix, where each row represents the vector of means $\bar{z}_i = \frac{1}{n}\sum_{j=1}^{n} z_{ij}$; h, the smoothing parameter.

Output: A scalar value.

`two.level.mvniw.BF`: Computes the BF for a two-level random effect model where both the within-source variability and the between-source variability are normally distributed, and the uncertainty about the within-source covariance matrix is modeled by an inverse Wishart distribution.

Usage: `two.level.mvniw.BF(quest,ref,O,B,mu,nw,p,n.iter, burn.in)`.

Arguments: quest, a ($n \times p$) matrix containing measurements on the questioned material; ref, a ($n \times p$) matrix containing measurements on the control material; O, the trace matrix of the inverse Wishart distribution; B, the between-source covariance matrix; mu, the mean vector between sources; nw, the number of degrees of freedom of the inverse Wishart distribution; p, the number of variables; n.iter, the number of iterations of the Gibbs sampling algorithm; burn.in, the number of discarded iterations.

Output: A scalar value.

two.level.mvniw2.BF: Computes the BF for a two-level random effect model where both the within-source variability and the between-source variability are normally distributed, the uncertainty about the within-source covariance matrix is modeled by an inverse Wishart distribution with no assumption of independence between questioned and known materials at the denominator (i.e., under H_2).

Usage: two.level.mvniw2.BF(quest,ref,Og,Bg,mug,Os,Bs,mus, nu,p,n.iter,burn.in).

Arguments: quest, a $(n \times p)$ matrix containing measurements on the questioned material; ref, a $(n \times p)$ matrix containing measurements on the control material; Og, the trace matrix of the inverse Wishart distribution from the database of genuine (handwritten) material; Bg, the between-source covariance matrix from the database of genuine (handwritten) material; mug, the mean vector between sources from the database of genuine (handwritten) material; Os, the trace matrix of the inverse Wishart distribution from the database of simulated (handwritten) material; Bs, the between-source covariance matrix from the database of simulated (handwritten) material; mus, the mean vector between sources from the database of simulated (handwritten) material; nw, the number of degrees of freedom of the inverse Wishart distribution; p, the number of variables; n.iter, the number of iterations of the Gibbs sampling algorithm; burn.in, the number of discarded iterations.

Output: A scalar value.

three.level.mv.WBV: Computes the estimate of the overall mean ϕ, the measurement error covariance matrix W, the within-group covariance matrix B, and the between-group covariance matrix V for the three-level model presented in Sect. 3.4.4.

Usage: three.level.mv.WBV(population,variables,grouping. item,grouping.fragment).

Arguments: population, a data frame with msn rows and k columns collecting measurements on m sources with s items for each source and n replicate measurements for each item; variables, a vector containing the column indices of the variables to be used; grouping.item, a scalar specifying the variable that is to be used as the grouping item; grouping.fragment, a scalar specifying the variable that is to be used for the grouping fragment.

Output: The estimated overall mean $\hat{\phi}$, the estimated measurement error covariance matrix $\hat{W}$, the estimated within-group covariance matrix $\hat{B}$, the estimated between-group covariance matrix $\hat{V}$.

three.level.mvn.BF: Computes the BF for a three-level random effect model where the variation at all three levels is normally distributed.

Usage: three.level.mvn.BF(bary,barx,barw,ny,nx,n1,n2,psi, W,B,V).

Arguments: bary, the mean vector of measurements on recovered items; barx, the mean vector of measurements on control items; barw, the mean vector of

measurements; ny, the number of recovered items; nx, the number of control items; n1, the number of replicate measurements on each of the recovered items; n2, the number of replicate measurements on each of the control items; psi, the overall mean vector; W, the replicate measurements covariance matrix; B, the within-group covariance matrix; V, the between-source covariance matrix.
Output: A scalar value.

Published with the support of the Swiss National Science Foundation (Grant no. 10BP12_208532/1).

Chapter 4
Bayes Factor for Investigative Purposes

4.1 Introduction

Forensic laboratories routinely face the problem of classifying items or individuals into one of several classes or populations on the basis of available data (e.g., measurements of one or more attributes), when no control material is available for comparison. As discussed in Sect. 1.6, forensic analyses can provide valuable information regarding the category membership of a particular item. For example, it may be of interest to classify banknotes seized on a person of interest as either banknotes from general circulation or banknotes related to drug trafficking (Wilson et al., 2014). The collected material is analyzed (e.g., the degree of contamination with cocaine is measured), and results are evaluated in terms of their effect on the odds in favor of a proposition H_1 according to which the recovered items originate from a given population (e.g., banknotes in general circulation), compared to an alternative proposition H_2 according to which the recovered items originate from another population (e.g., banknotes related to drug trafficking).

An assumption made throughout this chapter is that there is a finite number of populations to which an item of interest may belong. Each population will be characterized by a member from a family of probability distributions. Data can be either discrete or continuous, though for the latter it is easier to find examples and applications. There are many instances where the scientific evidence is described by several variables, and available measurements take the form of multivariate data. As mentioned in Sect. 3.1, data do not always present enough regularity so that standard parametric distributions could be used (e.g., the normal model). Moreover, data may present a complex dependence structure with several levels of variation.

Supplementary Information The online version contains supplementary material available at https://doi.org/10.1007/978-3-031-09839-0_4. The files can be accessed individually by clicking the DOI link in the accompanying figure caption or by scanning this link with the SN More Media App.

© The Author(s) 2022
S. Bozza et al., *Bayes Factors for Forensic Decision Analyses with R*,
Springer Texts in Statistics, https://doi.org/10.1007/978-3-031-09839-0_4

This chapter is structured as follows. Sections 4.2 and 4.3 address the problem of classification for various types of discrete and continuous data, respectively. Section 4.4 presents an extension to continuous multivariate data. Note that most of the examples developed in this chapter involve only two populations. An extension to more than two propositions is given in Sect. 4.2.2.

4.2   Discrete Data

This section deals with measurement results in the form of counts, using the binomial model (Sect. 4.2.1) and the multinomial model (Sect. 4.2.2).

4.2.1   Binomial Model

Imagine a case in which the issue is the quality of a consignment of Basmati rice. Basmati is a rice variety originating from the Indian subcontinent that became valuable in international trade in the last decades. This prompted the cultivation of high-yielding Basmati derivatives. Traditional and evolved (non-traditional) varieties, however, have distinct characteristics (e.g., Kamath et al., 2008), and distinguishing between varieties may be a relevant analytical task. Given a batch of Basmati rice of unknown type, the following pair of propositions may be of interest:

H_1:   The batch is traditional Basmati rice.
H_2:   The batch is non-traditional Basmati rice.

Denote by θ_1 and θ_2 the proportion of chalky grains in the two populations, respectively. Available counts can be treated as realizations of Bernoulli trials (Sect. 2.2.1) with constant probability of success θ_1 (θ_2). Suppose a conjugate beta prior distribution $\text{Be}(\alpha_i, \beta_i)$ is used to model uncertainty about θ_i, where α_i and β_i can be elicited using the available background knowledge (as in Sect. 1.10).

Among several characteristics of interest, such as grain length, thickness, weight, etc., is the percentage of chalky grains, determined by counting the number of grains having chalky area. A sample of size n is inspected, and a total number y of chalky grains are observed. This can be treated as a realization of a binomial distribution $\text{Bin}(n, \theta)$.

The marginal distribution at the numerator and denominator can be computed as in (1.25):

$$f_{H_i}(y) = \binom{n}{y} \frac{\Gamma(\alpha_i + \beta_i)\Gamma(\alpha_i + y)\Gamma(\beta_i + n - y)}{\Gamma(\alpha_i)\Gamma(\beta_i)\Gamma(\alpha_i + n + \beta_i)}.$$

This is a beta-binomial distribution with parameters n, α_i, and β_i. The Bayes factor in favor of proposition H_1 can be computed as in (1.26) and becomes

$$\frac{f_{H_1}(y)}{f_{H_2}(y)} = \frac{\Gamma(\alpha_1 + \beta_1)\Gamma(\alpha_1 + y)\Gamma(\beta_1 + n - y)\Gamma(\alpha_2)\Gamma(\beta_2)\Gamma(\alpha_2 + n + \beta_2)}{\Gamma(\alpha_2 + \beta_2)\Gamma(\alpha_2 + y)\Gamma(\beta_2 + n - y)\Gamma(\alpha_1)\Gamma(\beta_1)\Gamma(\alpha_1 + n + \beta_1)}. \quad (4.1)$$

Example 4.1 (Basmati Rice) Consider a case where 500 rice grains are examined and a total of 200 chalky grains are counted.

```
> n=500
> y=200
```

Suppose that the prior distribution for the proportion θ_1 of chalky grains in traditional varieties can be centered at 0.51 with a standard deviation equal to 0.19, while the proportion θ_2 of chalky grains in non-traditional varieties can be centered at 0.39 with a standard deviation equal to 0.31. The prior parameters (α_i, β_i) can be elicited as in (1.38) and (1.39).

```
> m1=0.51
> s1=0.19
> m2=0.39
> s2=0.31
```

We first write a function `beta_prior` that computes the prior parameters α_i and β_i according to (1.38) and (1.39).

```
> beta_prior=function(m,v){
+ a=m*(m*(1-m)/v-1)
+ b=(1-m)*(m*(1-m)/v-1)
+ return(c(a,b))}
```

The hyperparameters of the two beta distributions, say α_1, β_1, α_2, and β_2 can then be obtained straightforwardly as

```
> ab1=beta_prior(m1,s1^2)
> ab2=beta_prior(m2,s2^2)
```

The beta-binomial distribution can be calculated straightforwardly using the function `dbbinom` that is available in the package `extraDistr` (Wolodzko, 2020).

```
> library(extraDistr)
> BF=dbbinom(y,n,ab1[1],ab1[2])/dbbinom(y,n,ab2[1],
+ ab2[2])
> BF
```

```
[1] 2.009102
```

The Bayes factor provides weak support for the hypothesis that the rice type is traditional rather than non-traditional.

4.2.2   Multinomial Model

The physical and chemical analysis of gunshot residues (GSR) is a well-established field within forensic science. GSR are commonly analyzed to help with issues regarding the distance of firing and alleged activities of persons in incidents involving the use of firearms. A study by Brozek-Mucha and Jankowicz (2001) focused on the use of GSR for discriminating between a selected number of case types (i.e., particular combinations of weapon and ammunition). The authors conducted experiments using six categories, each consisting of a specific combination of weapon and ammunition, called categories A to F. Note that the aim here is not to infer a particular weapon and ammunition as the source of recovered GSR of unknown source. The purpose is only to provide assistance in discriminating between well-defined case types (i.e., categories).

Consider the following pair of competing propositions:

H_1:   The gunshot residue particles are of type D (Beretta pistol and 9 mm Luger ammunition).

H_2:   The gunshot residue particles are of type E (Margolin pistol with Sporting 5.6 mm ammunition).

Denote by θ_{1j} and θ_{2j} the proportion of particles in given chemical classes, $j = 1, \ldots, k$, characterizing categories D (i.e., category 1) and E (i.e., category 2). The number $n_1, \ldots, n_k$ of particles pertaining to distinct chemical classes $1, \ldots, k$, i.e., the chemical classes PbSbBa, PbSb, SbBa, Sb(Sn), Pb, and PbSnPb as specified in Brozek-Mucha and Jankowicz (2001), can be treated as realization from a multinomial distribution $f(n_1, \ldots, n_k \mid \theta_{i1}, \ldots, \theta_{ik})$, $i = 1, 2$. A conjugate Dirichlet prior probability distribution $f(\theta_{i1}, \ldots, \theta_{ik} \mid \alpha_{i1}, \ldots, \alpha_{ik})$ can be considered for modeling uncertainty about the proportions θ_{ij}, $i = 1, 2$ (Sect. 3.2.2).

The marginal distribution at the numerator and the denominator of the Bayes factor in (1.26) can be computed as in (1.25) and becomes

$$f_{H_i}(n_1, \ldots, n_k \mid \alpha_{i1}, \ldots, \alpha_{ik}) = \frac{\Gamma(\alpha_i)\Gamma(n+1)}{\Gamma(n+\alpha_i)} \prod_{j=1}^{k} \frac{\Gamma(n_k + \alpha_{ij})}{\Gamma(\alpha_{ij})\Gamma(n_j+1)},$$

where $\alpha_i = \sum_{j=1}^{k} \alpha_{ij}$ and $n = \sum_{j=1}^{k} n_j$. This is a Dirichlet-multinomial distribution with parameters n and $\alpha_{i1}, \ldots, \alpha_{ik}$.

From a decision-theoretic point of view, the questioned items can be classified in category D (decision d_1) whenever

$$\mathrm{BF} > \frac{l_1/l_2}{\pi_1/\pi_2}, \qquad (4.2)$$

where l_1 (l_2) represents the loss incurred when decision d_1 (d_2) is erroneous, and a "$0 - l_i$" loss function is chosen (Sect. 1.9 and Table 1.4), while π_1/π_2 is the prior odds in favor of H_1.

It may be objected that the values for l_1 and l_2 are difficult to assess. However, what really matters is the ratio k of the actual values, $l_1 = k \cdot l_2$. Note that this is an asymmetric loss function. In this way, starting from a prior odds equal to 1, the criterion in (4.2) may be rewritten as follows:

$$BF > k. \tag{4.3}$$

Stated otherwise, whenever the competing hypotheses are considered equally probable, a priori, the decision d_1 will be optimal if $BF > k$, that is if wrongly deciding d_1 (i.e., H_2 holds) is less than BF times worse than wrongly deciding d_2 (i.e., H_1 holds). Clearly, the prior odds must not necessarily be equal to 1, and the criterion can be adapted accordingly.

4.2.2.1 Choosing the Parameters of the Dirichlet Prior

The problem of how to elicit a prior probability distribution about a proportion has been discussed in Sect. 1.10. In the type of case considered here, an analyst will face the problem of eliciting a prior opinion about a set of proportions, assuming that the subjective prior distribution is chosen from the family of Dirichlet distributions.

There are various options for the hyperparameters $\alpha_{i1}, \ldots, \alpha_{ik}$, characterizing the prior probability distribution on the proportions $\theta_{i1}, \ldots, \theta_{ik}$. One is the uniform prior probability distribution, with $\alpha_{ij} = 1$, $j = 1, \ldots, k$. Whenever further information is available in terms of the number of outcomes in the distinct categories, e.g., $x_{i1}, \ldots, x_{ik}$, the hyperparameters α_{ij} can be updated to $\alpha_{ij} + x_{ij}$.

There are cases, however, where the analyst is able to specify a non-uniform prior probability distribution about the proportions. Following the methodology illustrated in Zapata-Vazquez et al. (2014), the prior probability distribution about a set of proportions $\theta_{i1}, \ldots, \theta_{ik}$ can be elicited using tools available in the package SHELF (Oakley, 2008). The user is only asked to provide a lower (e.g., 0.25), a median, and a upper (e.g., 0.75) quantile for the marginal densities of proportions that follow a beta distribution. Details will follow in the next example. The reader can also refer to O'Hagan et al. (2006), where a practical example is provided.

Example 4.2 (Gunshot Residue Particles) Consider a case in which a given number of particles (266) have been collected and analyzed by a scientist. The particles have been collected from a target surface (e.g., a person's hands). The counts of gunshot residue particles are as follows:

(continued)

Example 4.2 (continued)

Total number	Chemical classes					
of particles	PbSbBa	PbSb	SbBa	Sb(Sn)	Pb	PbSnPb
266	18	36	2	150	38	22

The scientist is asked to help discriminating between the following two propositions:

H_1: The gunshot residue particles are of type D (Beretta pistol with Luger 9 mm ammunition).

H_2: The gunshot residue particles are of type E (Margolin pistol with Sporting 5.6 mm ammunition).

One way to elicit the Dirichlet distribution in the case here is to use observed frequencies of particles in various chemical classes as reported in previous studies (e.g., Brozek-Mucha & Jankowicz, 2001). Suppose that the elicited expert judgments for the marginal proportions characterizing category D are as follows:

Quartiles	Chemical classes					
(%)	PbSbBa	PbSb	SbBa	Sb(Sn)	Pb	PbSnPb
Lower	5.00	9.00	0.40	66	9.00	7.60
Median	5.25	9.25	0.45	68	9.25	7.80
Upper	5.50	9.50	0.50	70	9.50	8.00

and those characterizing category E:

Quartiles	Chemical classes					
(%)	PbSbBa	PbSb	SbBa	Sb(Sn)	Pb	PbSnPb
Lower	2.35	7.00	0.13	56	24	5.60
Median	2.55	7.50	0.15	58	26	5.80
Upper	2.75	8.00	0.17	60	28	6.00

Consider, first, the elicitation of the Dirichlet distribution concerning the first population, $Dir(\theta_{11}, \ldots, \theta_{1k} \mid \alpha_{11}, \ldots, \alpha_{1k})$. Starting from the given lower, median, and upper quartiles for each marginal proportion, the prior distribution can be elicited as follows.

```
> p=c(0.25,0.5,0.75)
> th1=c(5,5.25,5.5)/100
```

(continued)

Example 4.2 (continued)
```
> th2=c(9,9.25,9.5)/100
> th3=c(0.4,0.45,0.5)/100
> th4=c(66,68,70)/100
> th5=c(9,9.25,9.5)/100
> th6=c(7.6,7.8,8)/100
```

The function `fitdist`, available in the package `SHELF`, allows one to fit a parametric distribution starting from the elicited probabilities. In the example here, the parameters of the elicited beta distribution for each proportion are of interest.

```
> library(SHELF)
> fit1=fitdist(vals = th1, probs = p, 0, 1)
> fit2=fitdist(vals = th2, probs = p, 0, 1)
> fit3=fitdist(vals = th3, probs = p, 0, 1)
> fit4=fitdist(vals = th4, probs = p, 0, 1)
> fit5=fitdist(vals = th5, probs = p, 0, 1)
> fit6=fitdist(vals = th6, probs = p, 0, 1)
```

The last six objects contain the parameters of the beta distribution that is fitted for each marginal proportion. For example, the parameters α_1 and β_1 of the elicited beta distribution of θ_1 (i.e., proportion of gunshot residue particles in category PbSbBa) can be obtained as

```
> fit1$Beta
```

```
    shape1   shape2
1 190.1306 3427.17
```

Next, fit the Dirichlet distribution to the elicited marginals by means of the function `fitDirichlet` that is available in the same package.

```
> d.fit = fitDirichlet(fit1,fit2,fit3,fit4,fit5,fit6,
+ categories = c("PbSbBa","PbSb","SbBa","Sb(Sn)",
+ "Pb","PbSnPb"),n.fitted = "min")
```

Directly elicited beta marginal distributions:

	PbSbBa	PbSb	SbBa	Sb(Sn)
shape1	1.90e+02	5.65e+02	3.67e+01	168.0000
shape2	3.43e+03	5.54e+03	8.06e+03	79.3000
mean	5.26e-02	9.25e-02	4.53e-03	0.6800
sd	3.71e-03	3.71e-03	7.46e-04	0.0296
sum	3.62e+03	6.11e+03	8.10e+03	248.0000

(continued)

Example 4.2 (continued)

```
              Pb     PbSnPb
shape1  5.65e+02  6.38e+02
shape2  5.54e+03  7.54e+03
mean    9.25e-02  7.80e-02
sd      3.71e-03  2.97e-03
sum     6.11e+03  8.18e+03


Sum of elicited marginal means: 1


Beta marginal distributions from Dirichlet fit:


          PbSbBa       PbSb      SbBa     Sb(Sn)
shape1   13.0000    22.9000  1.12e+00  168.0000
shape2  235.0000   225.0000  2.46e+02   79.3000
mean      0.0526     0.0925  4.53e-03    0.6800
sd        0.0142     0.0184  4.26e-03    0.0296
sum     248.0000   248.0000  2.48e+02  248.0000
              Pb    PbSnPb
shape1   22.9000   19.300
shape2  225.0000  228.000
mean      0.0925    0.078
sd        0.0184    0.017
sum     248.0000  248.000
```

The Dirichlet parameters $\alpha_{11}, \ldots, \alpha_{1k}$ can be read off from the row shape 1 and will be stored in a vector named a1.

```
> a1=c(13,22.9,1.12,168,22.9,19.3)
```

Parameter n of the Dirichlet prior is chosen by minimizing the sum of the beta parameters in each elicited marginal (input n.fitted set equal to min). See Oakley (2008) for more details.

In the same way, the Dirichlet distribution concerning the second population, $\text{Dir}(\theta_{21}, \ldots, \theta_{2k} \mid \alpha_{21}, \ldots, \alpha_{2k})$, can be elicited.

```
> th1=c(2.35,2.55,2.75)/100
> th2=c(7,7.5,8)/100
> th3=c(0.13,0.15,0.17)/100
> th4=c(56,58,60)/100
> th5=c(24,26,28)/100
> th6=c(5.6,5.8,6)/100
> fit1=fitdist(vals = th1, probs = p, 0, 1)
> fit2=fitdist(vals = th2, probs = p, 0, 1)
```

(continued)

Example 4.2 (continued)

```
> fit3=fitdist(vals = th3, probs = p, 0, 1)
> fit4=fitdist(vals = th4, probs = p, 0, 1)
> fit5=fitdist(vals = th5, probs = p, 0, 1)
> fit6=fitdist(vals = th6, probs = p, 0, 1)
> d.fit = fitDirichlet(fit1,fit2,fit3,fit4,fit5,fit6,
+ categories = c("PbSbBa","PbSb","SbBa","Sb(Sn)",
+ "Pb","PbSnPb"),n.fitted = "min")
```

The Dirichlet parameters $\alpha_{21}, \ldots, \alpha_{2k}$ can be read off analogously from the row shape 1 (not shown here) and will be stored in a vector named a2.

```
> a2=c(5.59,16.4,0.331,127,57,12.7)
```

The counts of gunshot residue particles are

```
> n=c(18,36,2,150,38,22)
```

The density of a Dirichlet-multinomial distribution can be calculated using the function ddirmnom that is available in the package extraDistr (Wolodzko, 2020), and the Bayes factor can be obtained straightforwardly

```
> library(extraDistr)
> BF=ddirmnom(n,sum(n),a1)/ddirmnom(n,sum(n),a2)
> BF
```

```
[1] 658.6326
```

The Bayes factor provides moderately strong support for the hypothesis that the gunshot residue particles originate from a Beretta pistol with Luger 9 mm ammunition rather than from a Margolin pistol with Sporting 5.6 mm ammunition.

Assume $\pi_1 = \pi_2 = 1$. If a "$0 - l_i$" loss function is introduced, then decision d_1, classifying the gunshot residue particles into category D, is to be preferred to the alternative decision d_2 unless wrongly deciding d_1 is felt more than 659 times worse than classifying the particles in category E.

Note that by choosing a "$0 - 1$" loss function, or a symmetric "$0 - l_i$" loss function with $l_1 = l_2$, a BF greater than 1 (or, more generally, greater than π_2/π_1 for unequal prior probabilities) provides a criterion for addressing the classification problem. The aim here was to show that when assuming equal prior probabilities for the hypotheses being compared, then, for the decision d_2 to be optimal, it is not sufficient to have an asymmetric loss function that assigns a loss to the adverse consequence of decision d_1 that is greater than the loss assigned to the adverse consequence of decision d_2. Specifically, this loss must be roughly 659 times greater.

4.2.2.2 More than Two Populations

Consider now the case where more than two weapons (and related ammunitions) could be at the origin of the collected gunshot particles. Suppose that a third weapon is taken into consideration and that the competing propositions are specified as follows:

H_1: The gunshot residue particles are of type D (Beretta pistol with Luger 9 mm ammunition; population p_1).

H_2: The gunshot residue particles are of type E (Margolin pistol with Sporting 5.6 mm ammunition; population p_2).

H_3: The gunshot residue particles are of type F (TT-33 pistol with Tokarev 7.62 mm ammunition; population p_3).

As discussed in Sect. 1.6, the expert may calculate the marginal likelihood $f_{H_i}(y)$ (i.e., a Dirichlet-multinomial distribution) for each proposition and report a scaled version as in (1.27), that is,

$$f_{H_i}^*(y) = \frac{f_{H_i}(y)}{\sum_{j=1}^3 f_{H_j}(y)},$$

or the posterior probabilities

$$\Pr(H_i \mid y) = \frac{\Pr(H_i) f_{H_i}^*(y)}{\sum_{j=1}^3 \Pr(H_j) f_{H_j}^*(y)}, \qquad i = 1, \ldots, 3.$$

Alternatively, the analyst may also consider the possibility of summarizing propositions H_2 and H_3 into one as $\bar{H}_1 = H_2 \cup H_3$. A pair of competing propositions may thus be formulated as follows:

H_1: The gunshot residue particles are of type D (Beretta pistol with Luger 9 mm ammunition; population p_1).

$\bar{H}_1$: The gunshot residue particles are of type E (Margolin pistol with Sporting 5.6 mm ammunition; population p_2) or of type F (TT-33 pistol with Tokarev 7.62 mm ammunition; population p_3).

The Bayes factor can be obtained as in (1.28), that is,

$$\mathrm{BF} = \frac{f_{H_1}(y) \sum_{i=2}^3 \Pr(p_i)}{f_{\bar{H}_1}(y)}, \qquad (4.4)$$

where

$$f_{\bar{H}_1}(y) = \sum_{i=2}^3 \Pr(p_i) \int_{\Theta_i} f(y \mid \theta_i) \pi(\theta_i \mid p_i) d\theta_i.$$

Example 4.3 (Gunshot Residue Particles—Continued) Recall Example 4.2, and suppose that the elicited expert judgments for the marginal propositions characterizing category F are as follows:

Quartiles	Chemical classes					
(%)	PbSbBa	PbSb	SbBa	Sb(Sn)	Pb	PbSnPb
Lower	6.00	4.50	3.00	65	14.0	3.00
Median	6.15	4.75	3.25	67	14.5	3.25
Upper	6.30	5.00	3.50	69	15.0	3.50

The Dirichlet distribution concerning this new combination of weapon/ammunition can be elicited as before:

```
> th1=c(6,6.15,6.30)/100
> th2=c(4.5,4.75,5)/100
> th3=c(3,3.25,3.5)/100
> th4=c(65,67,69)/100
> th5=c(14,14.5,15)/100
> th6=c(3,3.25,3.5)/100
> fit1=fitdist(vals = th1, probs = p, 0, 1)
> fit2=fitdist(vals = th2, probs = p, 0, 1)
> fit3=fitdist(vals = th3, probs = p, 0, 1)
> fit4=fitdist(vals = th4, probs = p, 0, 1)
> fit5=fitdist(vals = th5, probs = p, 0, 1)
> fit6=fitdist(vals = th6, probs = p, 0, 1)
> d.fit = fitDirichlet(fit1,fit2,fit3,fit4,fit5,fit6,
+ categories = c("PbSbBa","PbSb","SbBa","Sb(Sn)",
+ "Pb","PbSnPb"),n.fitted = "min")
```

The Dirichlet parameters $\alpha_{31}, \ldots, \alpha_{3k}$ can be read off from the row shape 1 (not shown here) and will be stored in a vector named a3.

```
> a3=c(15.7,12.1,8.29,170,36.9,8.29)
```

The scaled version of the marginal likelihoods can be easily obtained as

```
> fh1=ddirmnom(n,sum(n),a1)
> fh2=ddirmnom(n,sum(n),a2)
> fh3=ddirmnom(n,sum(n),a3)
> fh1scaled=fh1/(fh1+fh2+fh3)
> fh2scaled=fh2/(fh1+fh2+fh3)
```

(continued)

Example 4.3 (continued)
```
> fh3scaled=fh3/(fh1+fh2+fh3)
> c(fh1scaled,fh2scaled,fh3scaled)

[1]  0.9980356379  0.0015153146  0.0004490475
```

Note that the scaled likelihoods $f^*_{H_i}(y)$ are equivalent to the posterior probabilities $\Pr(H_i \mid y)$ whenever the prior probabilities of the three propositions are equal.

Alternatively, suppose that propositions H_2 and H_3 are summarized as above, i.e., $\bar{H}_1 = H_2 \cup H_3$, and that the prior probabilities of H_1 and $\bar{H}_1$ are equal, so that $\Pr(H_1) = 0.5$ and $\Pr(H_2) = \Pr(H_3) = 0.25$.

```
> p2=0.25
> p3=0.25
```

The Bayes factor can then be obtained as

```
> fh1=ddirmnom(n,sum(n),a1)
> fh2=p2*ddirmnom(n,sum(n),a2)+p3*
+ ddirmnom(n,sum(n),a3)
> BF=fh1*(p2+p3)/fh2
> BF

[1]  1016.142
```

4.3   Continuous Data

The previous section considered the evaluation of scientific evidence in the form of discrete data for investigative purposes. However, for many types of scientific evidence, measurements lead to continuous data. In this section, we discuss parametric and non-parametric models for continuous data.

4.3.1   Normal Model and Known Variance

Suppose that tablets of unknown source are seized, and the question is whether they belong to population A or population B, which differ in color dye concentration. The propositions of interest are as follows:

H_1:   The seized tablets come from population A.
H_2:   The seized tablets come from population B.

The measurement of color dye concentration leads to continuous data for which a normal distribution is considered appropriate, say $X_A \sim N(\theta_A, \sigma_A^2)$ for population A and $X_B \sim N(\theta_B, \sigma_B^2)$ for population B. Suppose that the variance of color dye concentration in the different populations is known. For the population means, a conjugate prior normal distribution is introduced, i.e., $\theta_A \sim N(\mu_A, \tau_A^2)$ and $\theta_B \sim N(\mu_B, \tau_B)$.

The analysis of a tablet of unknown origin yields the measurement y. The Bayes factor can be obtained as in (1.26), where the marginal likelihoods $f_{H_i}(y)$ are still normal with mean equal to the prior mean μ and variance equal to the sum of the prior variance τ^2 and the population variance σ^2, $f_{H_i}(y) = N(\mu, \tau^2 + \sigma^2)$.

Whenever several measurements $(y_1, \ldots, y_n)$ are available, it is sufficient to recall that the joint likelihood is proportional to the likelihood of the sample mean $\bar{y}$, which is normally distributed, $\bar{Y} \sim N(\theta, \sigma^2/n)$, and that the marginal likelihood in correspondence of the sample mean $\bar{y}$ becomes $f_{H_i}(\bar{y}) = N(\mu, \tau^2 + \sigma^2/n)$.

Example 4.4 (Color Dye Concentration in Ecstasy Tablets) A tablet of unknown origin is analyzed, and the measured color dye concentration is 0.16 (measurements are in %). A prior probability distribution is elicited for the mean of population A, as $\theta_A \sim N(0.14, 0.003^2)$, and for the mean of population B, as $\theta_B \sim N(0.3, 0.016^2)$. The population variances σ_A^2 and σ_B^2 are assumed to be known and equal to 0.01^2 and 0.06^2, respectively (Goldmann et al., 2004).

```
> y=0.160
> pma=0.14
> pva=0.003^2
> pmb=0.3
> pvb=0.016^2
> sigmaa=0.01^2
> sigmab=0.06^2
```

The Bayes factor in (1.26) can be obtained straightforwardly as the ratio of two normal likelihoods evaluated for the available measurement of color dye concentration y.

```
> BF=dnorm(y,pma,sqrt(pva+sigmaa))/
+ dnorm(y,pmb,sqrt(pvb+sigmab))
> BF

[1] 12.05706
```

The Bayes factor provides moderate support for the proposition according to which the analyzed tablet comes from population A, rather than the proposition according to which the tablet comes from population B. Note

(continued)

Example 4.4 (continued)

again that this result does not mean that proposition H_1 is more probable than proposition H_2. It solely means that the probability to observe the concentration y is roughly 12 times greater if the tablet originates from population A rather than from population B. The posterior odds might be in favor of proposition H_2 even in the presence of a Bayes factor greater than 1, if the prior probability of proposition H_1 is sufficiently small. In the case at hand, it can be easily verified that the prior probability of proposition H_1 needs to be smaller than 0.07 in order for the posterior odds to be in favor of H_2.

Suppose now that $n = 5$ tablets are available, and the color dye concentration measurements are $y = (0.155, 0.160, 0.165, 0.161, 0.159)$. The value of the evidence can then be computed for the sample mean

```
> y=c(0.155,0.160,0.165,0.161,0.159)
> n=length(y)
> num=dnorm(mean(y),pma,sqrt(pva+sigmaa/n))
> den=dnorm(mean(y),pmb,sqrt(pvb+sigmab/n))
> BF=num/den
> BF

[1] 134.628
```

The Bayes factor now provides moderately strong support for the proposition H_1, compared to proposition H_2. This is a direct effect of the increased number of measurements.

4.3.2   *Normal Model and Unknown Variance*

In some applications, both parameters are unknown, and a prior distribution for the population mean and the population variance must be introduced. A non-informative or a subjective prior distribution may be chosen, as mentioned previously in Sect. 3.3.2.

Consider a case where skeletal remains are analyzed, and the question is whether they belong to a man or a woman. The competing propositions are as follows:

H_1:   The skeletal remains belong to a woman.
H_2:   The skeletal remains belong to a man.

The study of Benazzi et al. (2009) found that the measurement of the sacral base is a useful indicator of sex.

Consider a normal probability distribution for the area of the sacral base $X_F \sim N(\theta_F, \sigma_F^2)$ for the population of females, and $X_M \sim N(\theta_M, \sigma_M^2)$ for the population of males. A conjugate prior probability distribution $f(\theta_i, \sigma_i^2)$ can be assumed for (θ_i, σ_i^2) as in (3.12), where $(\theta_i \mid \sigma_i^2) \sim N(\mu_i, \sigma_i^2/n_i)$ and $\sigma_i^2 \sim S_i \cdot \chi^{-2}(k_i)$, $i = \{F, M\}$. This amounts to an inverse gamma distribution with shape parameter $\alpha_i = k_i/2$ and scale parameter $\beta_i = S_i/2$, $\sigma_i^2 \sim IG(k_i/2, S_i/2)$.

The marginal density needed to compute the BF, $f_{H_i}(\cdot)$, is a Student t distribution with k_i degrees of freedom, centered at μ_i, with spread parameter, denoted here sp_i, equal to

$$sp_i = \frac{n_i}{n_i + 1} \alpha_i \beta_i^{-1}$$

(as noted previously in Sect. 3.3.2). Note that in this case there is one available measure ($n_y = 1$).

Example 4.5 (Sex Discrimination for Skeletal Remains) The sacral base of a skeletal remain is measured and found to be $11.5\,cm^2$. The prior probability distribution for (θ_p, σ_p^2), as illustrated in Sect. 3.3.2, is elicited based on the following population data:

Population	Females	Males
Number of individuals	38	35
Sample mean (cm^2)	10.35	14.09
Std dev (cm^2)	1.42	1.52

The prior distribution for $(\theta_F \mid \sigma_F^2)$ and $(\theta_M \mid \sigma_M^2)$ can be centered at $\mu_F = 10.35$ and $\mu_M = 14.09$, respectively, with $n_F = 38$ and $n_M = 35$.

```
> muf=10.35
> nf=38
> mum=14.09
> nm=35
```

The prior distribution for σ_F^2 and σ_M^2 can be elicited using the parameter value $k = 20$ (as in Example 3.6) and choosing S_F and S_M such that

$$Pr(\sigma_F^2 > 1.42^2) = Pr(\sigma_M^2 > 1.52^2) = 0.5$$

```
> k=20
> sigmaf=1.42^2
> sigmam=1.52^2
> q=qchisq(0.5,k)
```

(continued)

Example 4.5 (continued)
```
> Sf=q*sigmaf
> Sm=q*sigmam
> c(Sf,Sm)
```

```
[1] 38.99199 44.67720
```

The prior distributions for σ_F^2 and σ_M^2 are $39 \cdot \chi^{-2}(20)$ and $45 \cdot \chi^{-2}(20)$, respectively. The marginal density in the numerator of the Bayes factor is a Student t distribution with k_F degrees of freedom, centered at $\mu_F = 10.35$ with spread parameter $s_F = 0.5$ (rounded at the second decimal).

```
> spf=nf/(nf+1)*k/Sf
```

The marginal density in the denominator of the Bayes factor is a Student t distribution with k_M degrees of freedom, centered at $\mu_M = 14.09$ with spread parameter $s_M = 0.44$ (rounded at the second decimal).

```
> spm=nm/(nm+1)*k/Sm
```

Note that in this case $k_F = k_M = k$.

The density of a non-central Student t distributed random variable can be calculated using the function `dstp` available in the package `LaplacesDemon` (Hall et al., 2020). The Bayes factor can be obtained as follows:

```
> library(LaplacesDemon)
> y=11.5
> BF=dstp(y,muf,spf,k)/dstp(y,mum,spm,k)
> BF
```

```
[1] 3.184994
```

This value provides weak support for the proposition according to which the skeletal remains belong to a woman rather than a man.

4.3.3 Non-Normal Model

As pointed out in Sect. 3.4.1.2, certain types of observations lack sufficient regularity to apply standard parametric models.

Consider a case where banknotes are seized on an individual following an arrest. A question commonly asked in such a case is whether the seized banknotes come from a population of banknotes used in drug dealing activities. The following propositions may thus be formulated:

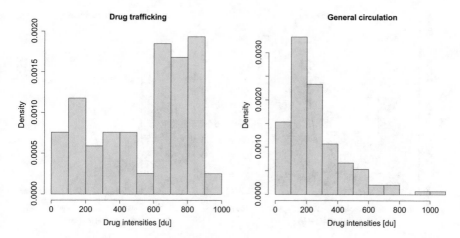

Fig. 4.1   Drug intensity measured on banknotes of 200 euro in a population of banknotes from drug trafficking (left) and general circulation (right) (Besson, 2004)

H_1:   The seized banknotes have been used in illegal drug dealing activities (population p_1).

H_2:   The seized banknotes are from general circulation (population p_2).

Figure 4.1 shows histograms of drug intensities measured on banknotes from drug trafficking (left) and general circulation (right). It can immediately be observed that the distributions for the two populations are different, that the distribution related to banknotes involved in drug trafficking is not unimodal, and that the one for banknotes in general circulation is positively skewed (Besson, 2004).

Suppose a database is available $\{\mathbf{z}_l = (z_{l1}, \ldots, z_{lm_l}), \; l = 1, 2\}$. The probability distribution for population p_l, $f_l(\cdot)$, can be estimated by means of kernel density estimation $\hat{f}_l(\cdot)$ as

$$\hat{f}_l(y \mid z_{l1}, \ldots, z_{lm_l}) = \frac{1}{m_l} \sum_{i=1}^{m_l} \mathrm{K}(y \mid z_{li}, h_l), \qquad (4.5)$$

where $\mathrm{K}(y \mid z_{li}, h_l)$ is taken to be a normal distribution centered at z_{li} with variance equal to $h_l^2 s_l^2$, $s_l^2 = \sum_{i=1}^{m_l} (z_{li} - \bar{z}_l)^2 / (m_l - 1)$, and $\bar{z}_l = \sum_{i=1}^{m_l} z_{li} / m_l$.

The estimate $\hat{f}_l(y)$ of the probability density is obtained by adding individual densities over all observations in the database and then dividing by the sum of the observations.

Figure 4.2 shows the kernel density estimates $\hat{f}_1(y \mid z_{11}, \ldots, z_{1m_1})$ and $\hat{f}_2(y \mid z_{21}, \ldots, z_{2m_2})$ obtained using (4.5) with the smoothing parameter set equal to 0.15 for both populations. It can be observed that kernel density estimates are more sensitive to multimodality and skewness and provide a better representation of the available data.

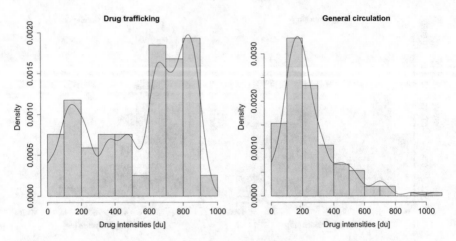

Fig. 4.2 Drug intensity measured on banknotes of 200 euro in a population of banknotes from drug trafficking (left) and general circulation (right), and associated kernel density estimates with smoothing parameter h equal to 0.15

Starting from the available measurements $y = (y_1, \ldots, y_n)$ on a sample of size n, a Bayes factor can be obtained as

$$\text{BF} = \frac{f_{H_1}(y)}{f_{H_2}(y)} = \frac{\prod_{i=1}^{m_1} \hat{f}_1(y_i \mid z_{11}, \ldots, z_{1m_1})}{\prod_{i=1}^{m_2} \hat{f}_2(y_i \mid z_{21}, \ldots, z_{2m_2})}. \tag{4.6}$$

Example 4.6 (Contaminated Banknotes) Consider a case in which 8 banknotes are seized on a person of interest. Laboratory analyses of the banknotes reveal drug intensities [du] equal to $y = (322, 158, 114, 125, 361, 801, 798, 135)$. A database named `banknotes.Rdata` is available on the book's website. It contains sample data for drug intensities on banknotes from drug trafficking and general circulation (Fig. 4.1). Note that these are hypothetical data used for the sole purpose of illustration. The $(n_1 \times 1)$ vector of measurements on banknotes from drug trafficking is extracted and denoted `pop1`; analogously, the $(n_2 \times 1)$ vector of measurements on banknotes from general circulation is extracted and denoted `pop2`.

```
> load('/.../banknotes.Rdata')
> po1=bancnotes[[1]]
> pop2=bancnotes[[2]]
```

(continued)

Example 4.6 (continued)

The smoothing parameters h_1 and h_2 are set equal to 0.15. The variances of drug concentration from each population, s_1^2 and s_2^2, are estimated by the sample variance

```
> h1=0.15
> h2=0.15
> s1=var(pop1)
> s2=var(pop2)
```

The kernel density estimation in (4.5) for the numerator and the denominator is computed by means of the functions kn1 and kn2, respectively.

```
> n1=length(pop1)
> n2=length(pop2)
> sk1=h1*sqrt(s1)
> sk2=h2*sqrt(s2)
> kn1=function(x){sum(dnorm(x,pop1,sk1))/n1}
> kn2=function(x){sum(dnorm(x,pop2,sk2))/n2}
```

The estimated probability densities are represented in Fig. 4.2.

```
> x=matrix(seq(0,1100,1),nrow=1)
> f1h=apply(x,2,kn1)
> f2h=apply(x,2,kn2)
> par(mfrow=c(1,2))
> hist(pop1,freq=F)
> lines(f1h,type='l')
> hist(pop2,freq=F)
> lines(f2h,type='l')
```

Consider now the vector of measurements y. The probability densities are estimated as in (4.5):

```
> y=matrix(c(322,158,114,125,361,801,798,135),nrow=1)
> f1=apply(y,2,kn1)
> f2=apply(y,2,kn2)
```

and the Bayes factor is obtained as in (4.6):

```
> BF=prod(f1)/prod(f2)
> BF
```

```
[1] 29.7187
```

The Bayes factor represents moderate support for the proposition according to which the seized banknotes have been used in illegal drug trafficking rather than the proposition according to which they are part of the general circulation.

Sensitivity to the Choice of the Smoothing Parameter

The sensitivity of the BF to the choice of the smoothing parameter may be a cause of concern, as different choices may be made. The smoothing parameter h determines the shape of the estimated probability density: if it is (too) large, the curve $\hat{f}(y)$ will be (very) smooth; on the other side, if it is (too) small, the resulting curve will be more spiky. Figure 4.3 shows, for both populations, the density curves obtained with $h = 0.1$ (dotted line), $h = 0.15$ (solid line), $h = 0.2$ (dashed line), $h = 0.25$ (dot-dashed line). The Bayes factor for the available measurements in Example 4.6 is then calculated for several choices of the smoothing parameter h.

```
> hsens=c(0.1,0.15,0.2,0.25)
> BFsens=rep(0,length(hsens))
> for (i in 1:length(hsens)){
+ sk1=hsens[i]*sqrt(s1)
+ sk2=hsens[i]*sqrt(s2)
+ f1=apply(y,2,kn1)
+ f2=apply(y,2,kn2)
+ BFsens[i]=prod(f1)/prod(f2)}
> round(BFsens,2)

[1] 1402.94    29.72     5.63     2.00
```

Note that the last two values correspond to large values of the smoothing parameter h, providing a very smooth curve.

4.4 Multivariate Data

As mentioned in Sect. 3.4, analysts frequently encounter multivariate data because the features of examined items and materials, such as handwritten or printed documents, glass fragments, or skeletal remains, can be described by more than one variable. Such data often present a complex dependence structure with a large number of variables and multiple levels of variation.

4.4.1 Normal Multivariate Data

The classification of skeletal remains on the basis of sexual dimorphism is a common problem in paleontology. Section 4.3.2 dealt with the question of how to quantify the evidential value of measurements of a given morphological trait (e.g.,

Drug trafficking

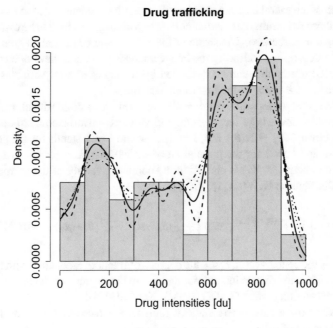

General circulation

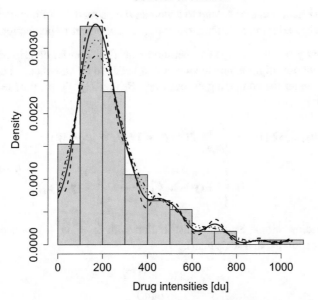

Fig. 4.3 Sample data used in Example 4.6 regarding drug intensities on banknotes for a population of banknotes from drug trafficking (top) and in general circulation (bottom), and associated kernel density estimates with smoothing parameter *h* equal to 0.1 (dashed line), 0.15 (solid line), 0.2 (dotted line), and 0.25 (dot-dashed line)

the profile of the sacral base). A number of studies have documented sex differences in particular pelvic traits, such as the *obturator foramen*, that tend to be oval in males and triangular in females. The shape of these traits can be described quantitatively by Fourier descriptors following the image analysis procedure developed by Bierry et al. (2010). Each item can be described by means of several variables, i.e., the amplitude and the phase of the first three harmonics.

Suppose that observations are available from a p-dimensional multivariate normal distribution whose mean vector and variance–covariance matrix are θ_l and W_l, respectively, $Z_{li} \sim N(\theta_l, W_l)$, $l = 1, 2$ (where $l = 1$ stands for the population of females and $l = 2$ for the population of males). Suppose further that the prior distribution about (θ_l, W_l) is chosen in the conjugate family of the normal-inverse Wishart distribution $NIW(\Omega_l, \nu_l, \mu_l, c_l)$:[1]

$$f(\theta_l, W_l) \propto |W_l|^{-(\nu_l+p+2)/2} \exp\left\{ -\frac{c_l}{2}(\theta_l - \mu_l)' W_l^{-1}(\theta_l - \mu_l) - \frac{1}{2}\mathrm{tr}(W_l^{-1}\Omega_l) \right\},$$

where μ_l is the center vector, c_l are the degrees of freedom associated with the center vector μ_l, Ω_l is the dispersion matrix, and ν_l are the degrees of freedom associated with the dispersion matrix Ω_l (O'Hagan & Kendall, 1994).

Consider now a case where skeletal remains are recovered, and the following propositions are of interest:

H_1:   The skeletal remains belong to a woman (i.e., a member of population p_1).
H_2:   The skeletal remains belong to a man (i.e., a member of population p_2).

Denote by $\mathbf{y} = (y_1, \dots, y_p)$ the measurements (i.e., Fourier descriptors) related to the item whose origin is unknown and that needs to be classified. The marginal distribution under the competing propositions H_1 and H_2, $f_{H_l}(\mathbf{y})$ for $l = 1, 2$, can be obtained as

$$f(\mathbf{y} \mid \mu_l, c_l, \Omega_l, \nu_l) = \int_{\theta_l, W_l} f(\mathbf{y} \mid \theta, W) f(\theta, W) d(\theta, W)$$

$$\propto \left\{ 1 + (\mathbf{y} - \mu_l)' \left[\frac{c_l+1}{c_l}\Omega_l \right]^{-1} (\mathbf{y} - \mu_l) \right\}^{-(\nu_l+1)/2} . \quad (4.7)$$

This is a p-dimensional Student t distribution with $\delta_l = \nu_l + 1 - p$ degrees of freedom, location μ_l, and scale matrix

$$\Delta_l = \frac{(c_l + 1)\Omega_l}{(c_l \delta_l)}.$$

---

[1] Note that a conjugate prior distribution may not always be the best choice. A method for assessing a non-conjugate prior distribution where the vector mean and the covariance matrix of the multivariate normal are, a priori, independent is provided by Garthwaite and Al-Awadhi (2001).

The Bayes factor can be obtained as

$$\mathrm{BF} = \frac{f(\mathbf{y} \mid \boldsymbol{\mu}_1, c_1, \Omega_1, \nu_1)}{f(\mathbf{y} \mid \boldsymbol{\mu}_2, c_2, \Omega_2, \nu_2)}.$$

4.4.1.1   Prior Distribution for the Unknown Mean and Variance

Four parameters must be elicited. The elicitation of $\boldsymbol{\mu}_l$ is rather simple. Since $\boldsymbol{\mu}_l$ represents the mean, the median, and the mode of the prior probability distribution, the analyst may assess any of these summaries (O'Hagan et al., 2006). A procedure for the elicitation of the degrees of freedom c and ν and the dispersion matrix Ω has been provided by Al-Awadhi and Garthwaite (1998).

Here, suppose a non-informative prior distribution is used:

$$f(\boldsymbol{\theta}_l, W_l) \propto \mid W_l \mid^{-(p+1)/2}.$$

A database is available, with n_1 measurements for the population of females (p_1) and n_2 measurements for the population of males (p_2). The corresponding posterior distributions (one for the numerator, one for the denominator) can be written as

$$(\boldsymbol{\theta}_l \mid \mathbf{z}_l, \Sigma_l) \sim \mathrm{N}(\bar{\mathbf{z}}_l, \Sigma_l / n_l) \tag{4.8}$$

$$(\Sigma_l \mid \mathbf{z}_l) \sim \mathrm{IW}(S_l, n_l - 1), \tag{4.9}$$

where $S_l = \sum_{i=1}^{n_l} (\mathbf{z}_{li} - \bar{\mathbf{z}}_l)(\mathbf{z}_{li} - \bar{\mathbf{z}}_l)'$ is the sum of the squares about the sample mean and $\bar{\mathbf{z}}_l = \sum_{j=1}^{n_l} \mathbf{z}_{lj} / n_l$.

The marginal likelihood $f_{H_l}(\mathbf{y})$ is, therefore, a p-dimensional Student t distribution with $n_l - p$ degrees of freedom, location vector $\bar{\mathbf{z}}_l$, and scale matrix

$$F_l = \frac{(n_l + 1)S_l}{n_l(n_l - p)}, \tag{4.10}$$

so that $(\mathbf{y} \mid \bar{\mathbf{z}}_l, F_l, n_l - p) \sim t_{n_l - p}(\bar{\mathbf{z}}_l, F_l)$.

Example 4.7 (Sex Discrimination for Skeletal Remains Using Multivariate Data)   Skeletal remains are recovered, and the obturator foramen area is measured. The measurements of the first three pairs of Fourier descriptors are as follows:

(continued)

Example 4.7 (continued)

First harmonic	Amplitude	0.083095
	Phase	2.6527709
Second harmonic	Amplitude	0.932333
	Phase	0.4530559
Third harmonic	Amplitude	0.413736
	Phase	0.3174581

Suppose that two databases of dimensions $(n_1 \times p) = (51 \times 6)$ and $(n_2 \times p) = (50 \times 6)$ are available for the population of women and men, respectively. These two databases can be used to obtain the summaries $\bar{z}_1, \bar{z}_2$ (i.e., the location vectors) and S_1, S_2 (i.e., the sum of the squares about the sample means) that are needed to calculate the marginal probability densities of the available measurements under the competing propositions. The location vectors z_1 and z_2 and the sum of the squares about the sample means S_1 and S_2 can be obtained straightforwardly as

```
> as.matrix(colMeans(population))
> cov(population)*(n-1)
```

where population is a database of dimension $(n \times p)$ containing the available data. Note that only summaries z_1, z_2, S_1, S_2, as well as the vector of measurements y are available in the database skeletal.Rdata and can be obtained as

```
> load('skeletal.Rdata')
> y

         A1        Phi1         A2        Phi2         A3
0.0830950 2.6527709 0.9323330 0.4530559 0.4137360
       Phi3
0.3174581

> cbind(m1,m2)

              [,1]         [,2]
A1     0.07500563 0.05078316
Phi1   2.60792515 3.37739963
A2     1.08366494 1.15684192
Phi2   0.17014670 0.08233948
A3     0.50490100 0.39364526
Phi3   0.34169629 0.39422141

> S1
```

(continued)

Example 4.7 (continued)

```
              A1         Phi1            A2
A1      0.09264018   0.3596888    0.16429701
Phi1    0.35968880  27.1073815    1.57099019
A2      0.16429701   1.5709902    2.34609053
Phi2    0.12045387   0.3766871   -0.03091293
A3      0.02310556  -0.7037211   -0.28908117
Phi3   -0.04738129   0.7762414   -0.36724194
              Phi2           A3          Phi3
A1      0.12045387   0.02310556  -0.04738129
Phi1    0.37668708  -0.70372110   0.77624136
A2     -0.03091293  -0.28908117  -0.36724194
Phi2    0.36278820  -0.02996462  -0.04588018
A3     -0.02996462   0.58676167   0.31452185
Phi3   -0.04588018   0.31452185   0.53788595

> S2

              A1          Phi1            A2
A1      0.059683655    0.41066454  -0.02342685
Phi1    0.410664544  138.15898708   2.29687413
A2     -0.023426848    2.29687413   1.53297489
Phi2    0.049798218   -1.91573412  -0.02475354
A3      0.082509024    0.01934154  -0.31589891
Phi3    0.007252672    2.47533633  -0.32754776
              Phi2           A3          Phi3
A1      0.04979822   0.08250902   0.007252672
Phi1   -1.91573412   0.01934154   2.475336335
A2     -0.02475354  -0.31589891  -0.327547756
Phi2    0.25584612   0.12310366  -0.149047658
A3      0.12310366   0.59361567   0.225155831
Phi3   -0.14904766   0.22515583   0.608557392
```

The marginal density $f_{H_1}(\mathbf{y})$ in the numerator of the Bayes factor is a p-dimensional Student t distribution with $n_1 - p = 45$ degrees of freedom, location m1 as above, and scale matrix

```
> n1=51
> p=6
> F1=S1*(n1+1)/(n1*(n1-p))
```

The marginal density $f_{H_2}(\mathbf{y})$ in the denominator of the Bayes factor is a p-dimensional Student t distribution with $n_2 - p = 44$ degrees of freedom, location m2 as above, and scale matrix

(continued)

Example 4.7 (continued)
```
> n2=50
> F2=S2*(n2+1)/(n2*(n2-p))
```

The density of a multivariate Student t distributed random variable can be calculated using the function `dmvt` available in the package `LaplacesDemon` (Hall et al., 2020).

```
> library(LaplacesDemon)
> num=dmvt(y,t(m1),F1,n1-p,log=FALSE)
> den=dmvt(y,t(m2),F2,n2-p,log=FALSE)
> num/den
```

```
[1] 1545.489
```

The Bayes factor represents strong support for the proposition according to which the skeletal remains originate from a woman (population p_1) rather than from a man (population p_2).

As discussed in Sect. 3.4.2, it is important to study the performance of the proposed model. This can be achieved by using the available databases to generate many test cases and computing relevant performance metrics.

4.4.1.2 Classification as a Decision

The BF obtained in Example 4.7 supports proposition H_1 over H_2. However, if a decision is to be made, one needs to take into account the prior uncertainty (in terms of probabilities) about the competing propositions and the undesirability (in terms of losses) of adverse outcomes (i.e., classification errors).

Let π_1 and π_2 denote the prior probabilities of propositions H_1 and H_2. The posterior probabilities α_1 and α_2 can be easily calculated as

$$\alpha_l = \frac{\pi_l f(\mathbf{y} \mid \boldsymbol{\mu}_l, c_l, \Omega_l, v_l)}{\sum_{j=1}^{2} \pi_j f(\mathbf{y} \mid \boldsymbol{\mu}_j, c_j, \Omega_j, v_j)},$$

where the marginals $f(\mathbf{y} \mid \boldsymbol{\mu}_j, c_j, \Omega_j, v_j), l = 1, 2$, are as in (4.7).

A criterion that can be used to classify the recovered item into one of the two populations has been outlined in Sect. 1.9. When using a "$0 - l_i$" loss function (Table 1.4), the Bayes decision criterion states that the decision d_1, classifying the recovered item in the population of females (p_1), is optimal whenever

$$\text{BF} > \frac{l_1/l_2}{\pi_1/\pi_2} = c. \tag{4.11}$$

Example 4.8 (Sex Discrimination for Skeletal Remains Using Multivariate Data—Continued) If the prior odds are 1, and a symmetric loss function is chosen (i.e., $l_1 = l_2$), the criterion in (4.11) says that the decision d_1 is optimal whenever BF > 1.

Assuming equal prior probabilities may be unrealistic because, often, there is at least some information to help assert whether one proposition is more probable than the stated alternative proposition. Likewise, the decision maker's preferences among adverse outcomes may not properly be reflected by a symmetric loss function, though it should be noted that what actually matters is only the ratio of l_1 to l_2.

To investigate the effect of alternative choices for the prior odds and the loss function, one can conduct a sensitivity analysis. Figure 4.4 shows an example for the threshold c in (4.11) as a function of increasing values of the prior probability π_1 and for different asymmetric loss functions, where l_2, the loss associated with the adverse outcome of the decision d_2, is fixed at 1, and l_1, associated with the adverse outcome of the decision d_1, is equal to 10, 50, and 100.

This analysis reveals that d_1 is not the optimal decision for very high values of l_1, compared to l_2, and for very small values of the prior probability π_1.

Fig. 4.4 Threshold c, the BF necessary to ensure that the decision d_1 has the smaller expected loss than the decision d_2, as specified by Eq. (4.11), as a function of the prior probability π_1 and for different loss ratios l_1 / l_2)

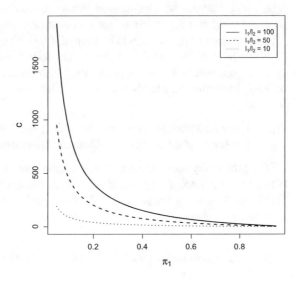

4.4.2   Two-Level Models

A recurrent problem in forensic practice is to help distinguish between legal and illegal cannabis plants (Bozza et al., 2014). Cannabis seedlings can be discriminated, to some extent, on the basis of their chemical profiles using chemometric tools and a methodology as described in Broséus et al. (2010). This study focused on several target compounds, taking into account their presence in drug type (illegal) and fiber type (legal) Cannabis.

Suppose a dataset is available that consists of replicate measurements (n) made on illegal plants (population p_1) and on fiber type plants (population p_2). The sample size is equal to m_1 and m_2 for populations p_1 and p_2, respectively. Background data can be denoted by $\mathbf{z}_{lij} = (z_{lij1}, \ldots, z_{lijp})$, where $l = 1, 2$, $i = 1, \ldots, m_l$, $j = 1, \ldots, n$, and p is the number of variables. Available data suggest that a statistical model with two levels of variation is suitable: variation between replicate measurements from the same source and variation between measurements from different sources.

4.4.2.1   Normal Distribution for the Between-Source Variation

Here we use the two-level random effect model described in Sect. 3.4.1.1. For the within-source variation, the distribution of Z_{lij} is taken to be normal, $Z_{lij} \sim N(\boldsymbol{\theta}_{li}, W_l)$. For the between-source variation, denote the mean vector between sources by $\boldsymbol{\mu}_l$, and the matrix of between-source variances and covariances by B_l. The distribution of $\boldsymbol{\theta}_{li}$ is taken to be normal, $\boldsymbol{\theta}_{li} \sim N(\boldsymbol{\mu}_l, B_l)$.

Measurements are available on some seized material, denoted by $\mathbf{y} = (\mathbf{y}_1, \ldots, \mathbf{y}_n)$, where $\mathbf{y}_j = (\mathbf{y}_{j1}, \ldots, \mathbf{y}_{jp})$, $j = 1, \ldots, n$. A laboratory is asked to help determine the plant's chemotype. The following propositions may be of interest:

H_1:   The seized plant is drug type Cannabis (population p_1).
H_2:   The seized plant is fiber type Cannabis (population p_2).

The probability distribution of the measurements on items from each population is taken to be normal, $Y \sim N(\boldsymbol{\theta}_l, B_l), l = 1, 2$. The marginal probability densities in the numerator and denominator have the form $f_{H_l}(\mathbf{y}) = f_l(\mathbf{y} \mid \boldsymbol{\mu}_l, W_l, B_l), l = 1, 2$, and can be obtained as in (3.28)

$$f_l(\mathbf{y} \mid \boldsymbol{\mu}_l, W_l, B_l) = \mid 2\pi W_l \mid^{-n/2} \mid 2\pi B_l \mid^{-1/2} \mid 2\pi (nW_l^{-1} + B_l^{-1})^{-1} \mid^{1/2}$$

$$\times \exp\left\{ -\frac{1}{2}\left[(\bar{\mathbf{y}} - \boldsymbol{\mu}_l)'(n^{-1}W_l + B_l)^{-1}(\bar{\mathbf{y}} - \boldsymbol{\mu}_l) + \mathrm{tr}\left(SW_l^{-1} \right) \right] \right\}, \qquad (4.12)$$

where $S = \sum_{i=1}^{n}(\mathbf{y}_i - \bar{\mathbf{y}})(\mathbf{y}_i - \bar{\mathbf{y}})'$.

The Bayes factor can then be obtained as in (1.26) as a ratio between the two marginals

$$BF = \frac{f_{H_1}(\mathbf{y})}{f_{H_2}(\mathbf{y})} = \frac{f_1(\mathbf{y} \mid, \boldsymbol{\mu}_1, W_1, B_1)}{f_2(\mathbf{y} \mid \boldsymbol{\mu}_2, W_2, B_2)}$$

$$= \left(\frac{|W_1|}{|W_2|}\right)^{-\frac{n}{2}} \left(\frac{|B_1|}{|B_2|}\right)^{-\frac{1}{2}} \left(\frac{\left| \left(n W_1^{-1} + B_1^{-1}\right)^{-1} \right|}{\left| \left(n W_2^{-1} + B_2^{-1}\right)^{-1} \right|}\right)^{\frac{1}{2}}$$

$$\times \exp\left\{\sum_{i=1}^{2}(-1)^i \frac{1}{2}\left[\operatorname{tr}(SW_i)^{-1} + (\bar{\mathbf{y}} - \boldsymbol{\mu}_i)' \left(n^{-1}W_i + B_i\right)^{-1}(\bar{\mathbf{y}} - \boldsymbol{\mu}_i)\right]\right\}.$$

$$(4.13)$$

The overall means $\boldsymbol{\mu}_1$ and $\boldsymbol{\mu}_2$, the within-source covariance matrices W_1 and W_2, and the between-source covariance matrices B_1 and B_2 can be estimated from the available background data using (3.32), (3.33), and (3.34).

Example 4.9 (Cannabis Seedlings) A plant of unknown type is analyzed, and the chemical profile is extracted. Three replicate measurements are taken ($n = 3$) on three variables ($p = 3$): Cannabidiol (CBD), D9-Tetrahydrocannabinol (THC), and Cannabinol (CBN). Measurements on the item of unknown type are as follows:

CBD	THC	CBN
−1.3040	0.2310	0.6874
−1.2918	0.2400	0.7350
−1.0719	0.3176	0.9113

```
> y=matrix(c(-1.304,0.231,0.6874,-1.2918,0.24,0.735,
+ -1.0710,0.3176,0.9113),nrow=3,byrow=T)
```

The mean vectors between sources $\boldsymbol{\mu}$, the within-source covariance matrices W, and the between-source covariance matrices B can be estimated from the available background data (Bozza et al., 2014).

The estimates of the overall means $\boldsymbol{\mu}_1$ and $\boldsymbol{\mu}_2$ of the within-source covariance matrices W_1 and W_2 and of the between-source covariance matrices B_1 and B_2 are available in the database plant.Rdata and can be obtained as

(continued)

Example 4.9 (continued)
```
> load('plant.Rdata')
> mu1
```

```
          CBD        THC        CBN
[1,] -0.4566709 0.9728053 0.9196972
```

```
> mu2
```

```
         CBD         THC         CBN
[1,] 0.4097014 -0.7850832 -0.7592971
```

```
> W1
```

```
          CBD          THC          CBN
CBD 0.01995126 0.015787374 0.010380235
THC 0.01578737 0.015708590 0.005226694
CBN 0.01038024 0.005226694 0.094354823
```

```
> W2
```

```
              CBD          THC           CBN
CBD  0.0180694402 1.901708e-03 -3.699212e-04
THC  0.0019017082 5.685754e-04  7.930402e-05
CBN -0.0003699212 7.930402e-05  1.878924e-02
```

```
> B1
```

```
          CBD       THC       CBN
CBD 0.4154039 0.2135218 0.1470832
THC 0.2135218 0.4752159 0.3893965
CBN 0.1470832 0.3893965 0.4292913
```

```
> B2
```

```
           CBD        THC        CBN
CBD 1.10811258 0.05630523 0.01847022
THC 0.05630523 0.06703743 0.05462002
CBN 0.01847022 0.05462002 0.10964122
```

These estimates can be obtained using the function `two.level.mv.WB` introduced in Sect. 3.4.1.1

```
> two.level.mv.WB(population,variables,
+ grouping.object)
```

where population is a data frame with the available data, variables indicates the columns where variables are displayed, and grouping.object indicates the item number.

(continued)

Example 4.9 (continued)

Given the available measurements, the Bayes factor can be calculated as in (4.13) using the function `two.level.mvn.inv.BF`.

```
> BF=two.level.mvn.inv.BF(y,W1,W2,B1,B2,mu1,mu2)
> BF

[1] 48739.7
```

The Bayes factor represents very strong support for the proposition according to which the seized plant is of drug type rather than fiber type.

4.4.2.2 Non-normal Distribution for the Between-Source Variation

As noted in Sect. 3.4.1.2, whenever the assumption of normality for the between-source variability is considered inappropriate, the normal distribution $f(\theta_{li} \mid \mu_l, B_l) = \mathrm{N}(\mu_l, B_l)$ previously proposed can be replaced by a kernel density estimate as in (3.35). The marginal densities $f_{H_i}(y)$ at the numerator and denominator of the Bayes factor become

$$f_l(\bar{\mathbf{y}} \mid W_l, B_l, h_l) = (2\pi)^{-p} \mid B_l \mid^{-1} (m_l h_l^2)^{-2} \mid D_l \mid^{-1/2} \mid D_l^{-1} + (h_l^2 B_l)^{-1} \mid^{-1/2}$$

$$\times \sum_{i=1}^{m_l} \exp\left\{ -\frac{1}{2}(\bar{\mathbf{y}} - \bar{\mathbf{z}}_{li})'(D_l + h_l^2 B_l)^{-1}(\bar{\mathbf{y}} - \bar{\mathbf{z}}_{li}) \right\}, \qquad (4.14)$$

where $D_l = n^{-1} W_l$. Note that this is just the marginal density of the recovered data, that is, the first line in (3.38), with all multiplicative constants.

The Bayes factor is then given by the ratio of the marginal probability densities in (4.14) for $l = 1, 2$, that is,

$$\mathrm{BF} = \frac{f_1(\bar{\mathbf{y}} \mid W_1, B_1, h_1)}{f_2(\bar{\mathbf{y}} \mid W_2, B_2, h_2)}. \qquad (4.15)$$

Example 4.10 (Cannabis Seedlings—Continued) Consider again the case examined in Example 4.9, and suppose that a kernel distribution is used to model the between-source variability. First, the group means $\bar{\mathbf{z}}_{li}$ must be obtained. They can be obtained as an output of the function `two.level.mv.WB` that can be used to estimate the model parameters.

(continued)

Example 4.10 (continued)
```
> head(group.means.1)

          CBD       THC       CBN
1 -1.22249231  0.2629209  0.777929
2 -0.04734919  1.7607730  2.293862
3 -0.59036072  1.1574978  1.403290
4 -0.27733591  1.5211215  1.832527
5 -0.54204482  1.2387804  1.545526
6 -0.65989575 -0.9686288  1.831042

> head(group.means.2)

            CBD        THC        CBN
141 -0.12963445 -1.0232887 -0.896759
142 -0.16827410 -0.9934113 -0.896759
143 -0.61568550 -1.0464456 -0.896759
144  0.03267767 -0.9815586 -0.896759
145  0.12647601 -0.9349308 -0.896759
146 -0.51730995 -0.9909842 -0.896759

> m1=dim(group.means.1)[1]
> m2=dim(group.means.2)[1]
> c(m1,m2)

[1] 117 155
```

Here we show only the first six rows of the $(m_l \times p)$ matrices, where each row represents the vector of means $\bar{z}_{li} = \frac{1}{n} \sum_{j=1}^{n} z_{lij}, l = 1, 2$. Note that the group means $\bar{z}_1$ and $\bar{z}_2$, as well as all the estimated parameters (μ_1, μ_2, W_1, W_2, B_1 and B_2) are available in the database plant.Rdata.

The smoothing parameters h_1 and h_2 in the two populations can be estimated as in (3.36), using the function hopt:

```
> p=3
> h1=hopt(p,m1)
> h2=hopt(p,m2)
> c(h1,h2)

[1] 0.4675469 0.4491338
```

Given the available measurements, the Bayes factor can be calculated as in (4.15) using the function two.level.mvk.inv.BF available in the supplementary materials available on the book's website

```
> BF=two.level.mvk.inv.BF(y,group.means.1,
+ group.means.2,W1,W2,B1,B2,h1,h2)
```

(continued)

> *Example 4.10* (continued)
> ```
> > BF
> [1] 7.42
> ```
>
> The Bayes factor represents moderate support for the proposition according to which the seized plant is drug type Cannabis rather than fiber type Cannabis.

4.4.2.3 Assessing Model Performance

One way to investigate the performance of the two models described in Sects. 4.4.2.1 and 4.4.2.2, denoted here Model 1 and Model 2, is to calculate a Bayes factor for all available measurements on items from population 1 (drug type). One would expect to obtain BFs greater than 1 (see Table 4.1). Clearly, one should also consider BF computations for all measurements on items from population p_2 (fiber type). In the latter case, BFs smaller than 1 would be expected (see Table 4.2).

Table 4.1 Bayes factor values for items of population 1 (Example 4.9 and 4.10) obtained using (4.13) (Method 1) and (4.15) (Method 2)

BF	Model 1	Model 2
$< 10^{-1}$	2	2
$10^{-1} - 1$	1	3
$1 - 10$	2	7
$10 - 10^2$	0	7
$10^2 - 10^3$	2	9
$10^3 - 10^4$	3	8
$10^4 - 10^5$	5	2
$10^5 - 10^6$	4	3
$10^6 - 10^7$	1	5
$10^7 - 10^8$	6	3
$10^8 - 10^9$	1	4
$10^9 - 10^{10}$	8	2
$10^{10} - 10^{100}$	82	62
Number of BFs > 1	114	112
Number of BFs < 1	3	5

Table 4.2 Bayes factor values for items of population 2 (Example 4.9 and 4.10) obtained using (4.13) (Method 1) and (4.15) (Method 2)

BF	Model 1	Model 2
$< 10^{-10}$	20	85
$10^{-10} - 10^{-9}$	4	0
$10^{-9} - 10^{-8}$	8	0
$10^{-8} - 10^{-7}$	10	0
$10^{-7} - 10^{-6}$	14	0
$10^{-6} - 10^{-5}$	29	0
$10^{-5} - 10^{-4}$	19	0
$10^{-4} - 10^{-3}$	20	0
$10^{-3} - 10^{-2}$	16	35
$10^{-2} - 10^{-1}$	2	23
$10^{-1} - 1$	1	7
$1 - 10$	6	3
$10 - 10^2$	0	1
$10^2 - 10^3$	0	1
$> 10^4$	6	0
Number of BFs > 1	12	5
Number of BFs < 1	143	150

4.5   Summary of R Functions

The R functions outlined below have been used in this chapter.

Functions Available in the Base Package

apply: Applies a function to the margins (either rows or columns) of a matrix.

colMeans: Forms column means for numeric arrays (or data frames).

d<name of distribution> (e.g., dnorm): Calculates the density for many parametric distributions.

More details can be found in the Help menu, help.start().

Functions Available in Other Packages

dbbinom and ddirmnom in the package extraDistr: Calculate the density of a beta-binomial distribution and that of a Dirichlet-multinomial distribution, respectively.

dstp and dmvt in the package LaplacesDemon: Calculate the density of a non-central Student t distribution and of a non-central multivariate Student t distribution, respectively.

fitdist and fitDirichlet in the package SHELF: Fit a parametric distribution starting from elicited probabilities and a Dirichlet distribution from the elicited beta distributions for a set of proportions, respectively.

Functions Developed in the Chapter

beta_prior: Calculates the hyperparameters α and β of a beta distribution $\text{Be}(\alpha, \beta)$ starting from the prior mean m and the prior variance v.
Usage: beta_prior(m,v).
Arguments: m, the prior mean; v, the prior variance.
Output: A vector of values, the first is α, the second is β.

hopt: Calculates the estimates $\hat{h}$ of the smoothing parameter h.
Usage: hopt(p,m).
Arguments: p, the number of variables; m, the number of sources.
Output: A scalar value.

kn1: Computes the kernel density estimation (numerator).
Usage: kn1(x,pop1,sk1).
Arguments: x, a vector of available measurements; pop1, a vector of measurements of drug intensities on banknotes from drug trafficking where the kernel is centered; sk1, the variance $h_1^2 s_1^2$ of the kernel, where h_1 is the smoothing parameter and s_1^2 is the sample variance of the available measurements.
Output: A scalar value.

post_distr: Computes the posterior distribution $N(\mu_x, \tau_x^2)$ of a normal mean θ, with $X \sim N(\theta, \sigma^2)$ and $\theta \sim N(\mu, \tau^2)$.
Usage: post_distr(sigma,n,barx,pm,pv).
Arguments: sigma, the variance σ^2 of the observations; n, the number of observations; barx, the sample mean $\bar{x}$ of the observations. pm, the mean μ of the prior distribution $N(\mu, \tau^2)$; pv, the variance τ^2 of the prior distribution $N(\mu, \tau^2)$.
Output: A vector of two values, the first is the posterior mean μ_x, the second is the posterior variance τ_x^2.

two.level.mv.WB: Computes the estimate of the overall mean $\boldsymbol{\mu}$, the group means $\bar{\mathbf{z}}_i$, the within-group covariance matrix W, and the between-group covariance matrix B.
Usage: two.level.mv.WB(population, variables,grouping.variable)
Arguments: population, a data frame with N rows and k columns collecting measurements on m sources with n_i items for each source, $i = 1, \ldots, m$; variables, a vector containing the column indices of the variables to be used; grouping.variable, a scalar specifying the variable that is to be used as the grouping factor.
Output: The group means $\bar{\mathbf{z}}_i$, the estimated overall mean $\hat{\boldsymbol{\mu}}$, the estimated within-group covariance matrix $\hat{W}$, the estimated between-group covariance matrix $\hat{B}$.

two.level.mvn.inv.BF: Computes the BF for investigative purposes from a two-level model where both the within-source variability and the between-source variability are normally distributed.
Usage: two.level.mvn.inv.BF(y,W1,W2,B1,B2,mu1,mu2, variables).

Arguments: y, a ($n \times p$) matrix of measurements; W$_1$ and W$_2$, the within-source covariance matrices; B$_1$ and B$_2$, the between-source covariance matrices; the overall group means μ_1 and μ_2; variables, a vector containing the column indices of the variables to be used.

Output: A scalar value.

two.level.mvk.inv.BF: Computes the BF for investigative purposes from a two-level model where the within-source variability is assumed to be normally distributed, while the between-source variability is modeled by a kernel density.

Usage: two.level.mvk.inv.BF(y,gmu1,gmu2,W1,W2,B1,B2,h1,h2).

Arguments: y, a ($n \times p$) matrix of measurements; gmu1 and gmu2, the group means $\bar{z}_{1i}$ and $\bar{z}_{2i}$; W$_1$ and W$_2$, the within-source covariance matrices; B$_1$ and B$_2$, the between-source covariance matrices; h$_1$ and h$_2$, the smoothing parameters h$_1$ and h$_2$.

Output: A scalar value.

Published with the support of the Swiss National Science Foundation (Grant no. 10BP12_208532/1).

References

Aitken, C. G. G. (1999). Sampling - How big a sample? *Journal of Forensic Sciences, 44*, 750–760.

Aitken, C. G. G., & Gold, E. (2013). Evidence evaluation for discrete data. *Forensic Science International, 230*, 147–155.

Aitken, C. G. G., & Lucy, D. (2004). Evaluation of trace evidence in the form of multivariate data. *Applied Statistics, 53*, 109–122. Supplementary materials (data) available at https://rss.onlinelibrary.wiley.com/hub/journal/14679876/series-c-datasets/pre_2016

Aitken, C. G. G., & Taroni, F. (2021). The history of forensic inference and statistics: A thematic perspective. In D. Banks, K. Kadafar, D. H. Kaye, & M. Tackett (Eds.), *Handbook of forensic statistics* (pp. 3–36). Boca Raton: CRC Press.

Aitken, C. G. G., Lucy, D., Zadora, G., & Curran, J. M. (2006). Evaluation of transfer evidence for three-level multivariate data with the use of graphical models. *Computational Statistics & Data Analysis, 50*, 2571–2588.

Aitken, C. G. G., Roberts, P., & Jackson, G. (2010). *Fundamentals of Probability and Statistical Evidence in Criminal Proceedings* (Practitioner Guide No. 1), Guidance for Judges, Lawyers, Forensic Scientists and Expert Witnesses, Royal Statistical Society's Working Group on Statistics and the Law.

Aitken, C. G. G., Taroni, F., & Bozza, S. (2021). *Statistics and the evaluation of evidence for forensic scientists* (3rd ed.). Chichester: Wiley, Chichester.

Al-Awadhi, S. A., & Garthwaite, P. H. (1998). An elicitation method for multivariate normal distributions. *Communications in Statistics - Theory and Methods, 27*, 1123–1142.

Albert, J. (2009). *Bayesian computation with R* (2nd ed.). Dordrecht: Springer.

Association of Forensic Science Providers (2009). Standards for the formulation of evaluative forensic science expert opinion. *Science & Justice, 49*, 161–164.

Balding, D. J., & Nichols, R. A. (1994). DNA profile match probability calculation: How to allow for population stratification, relatedness, database selection and single bands. *Forensic Science International, 64*, 125–140.

Benazzi, S., Maestri, C., Parisini, S., Vecchi, F., & Gruppioni, G. (2009). Sex assessment from the sacral base by means of image processing. *Journal of Forensic Sciences, 54*, 249–254.

Berger, J., & Pericchi, L. (2015). Bayes factors. *Wiley StatsRef: Statistics Reference Online* (pp. 1–14)

Berger, J. O. (1985). *Statistical decision theory and Bayesian analysis* (2nd edn.). New York: Springer.

Bernardo, J. M., & Smith, A. F. M. (2000). *Bayesian theory* (2nd edn.). Chichester: Wiley.

Besson, L. (2004). Détection des stupéfiants par IMS. Technical report, Institut de police scientifique, Université de Lausanne.

© The Author(s) 2022

S. Bozza et al., *Bayes Factors for Forensic Decision Analyses with R*,
Springer Texts in Statistics, https://doi.org/10.1007/978-3-031-09839-0

Biedermann, A., Taroni, F., Bozza, S., & Aitken, C. G. G. (2008). Analysis of sampling issues using Bayesian networks. *Law, Probability & Risk, 7*, 35–60.

Biedermann, A., Taroni, F., Bozza, S., & Mazzella, W. D. (2009). Implementing statistical learning methods through Bayesian networks (Part 1): A guide to Bayesian parameter estimation using forensic science data. *Forensic Science International, 193*, 63–71.

Biedermann, A., Taroni, F., Bozza, S., & Mazzella, W. D. (2011a). Implementing statistical learning methods through Bayesian networks (Part 2): Bayesian evaluations for results of black toner analyses in forensic document examination. *Forensic Science International, 204*, 58–66.

Biedermann, A., Bozza, S., & Taroni, F. (2011b). Probabilistic evidential assessment of gunshot residue particle evidence (Part II): Bayesian parameter estimation for experimental count data. *Forensic Science International, 206*, 103–110.

Biedermann, A., Bozza, S., Garbolino, P., & Taroni, F. (2012). Decision-theoretic analysis of forensic sampling criteria using Bayesian decision networks. *Forensic Science International, 223*, 217–227.

Biedermann, A., Garbolino, P., & Taroni, F. (2013). The subjectivist interpretation of probability and the problem of individualization in forensic science. *Science & Justice, 53*, 192–200.

Biedermann, A., Bozza, S., & Taroni, F. (2015). Prediction in forensic science: A critical examination of common understandings. *Frontiers in Psychology, 6*, 1–4.

Biedermann, A., Bozza, S., Taroni, F., Fürbach, M., Li, B., & Mazzella, W. (2016a). Analysis and evaluation of magnetism of black toners on documents printed by electrophotographic systems. *Forensic Science International, 267*, 157–165.

Biedermann, A., Bozza, S., & Taroni, F. (2016b). The decisionalization of individualization. *Forensic Science International, 266*, 29–38.

Biedermann, A., Bozza, S., Taroni, F., & Aitken, C. G. G. (2017a). The consequences of understanding expert probability reporting as a decision. *Science & Justice, 57*, 80–483. Special Issue on Measuring and Reporting the Precision of Forensic Likelihood Ratios.

Biedermann, A., Bozza, S., Taroni, F., & Aitken, C. G. G. (2017b). The meaning of justified subjectivism and its role in the reconciliation of recent disagreements over forensic probabilism. *Science & Justice, 57*, 477–483.

Biedermann, A., Taroni, F., Bozza, S., Augsburger, M., & Aitken, C. G. G. (2018). Critical analysis of forensic cut-offs and legal thresholds: A coherent approach to inference and decision. *Forensic Science International, 288*, 72–80.

Bierry, G., Le Minor, J. M., & Schmittbuhl, M. (2010). Oval in males and triangular in females? A quantitative evaluation of sexual dimorphism in the human obturator foramen. *American Journal of Phisical Anthropology, 141*, 626–631.

Bolck, A., Ni, H., & Lopatka, M. (2015). Evaluating score- and feature-based likelihood ratio models for multivariate continuous data. *Law, Probability & Risk, 14*, 243–266.

Bolstad, W. M., & Curran, J. M. (2017). *Introduction to Bayesian statistics* (3rd ed.). Hoboken: Wiley.

Bozza, S., Taroni, F., Marquis, R., & Schmittbuhl, M. (2008). Probabilistic evaluation of handwriting evidence: likelihood ratio for authorship. *Applied Statistics, 57*, 329–341.

Bozza, S., Broséus, J., Esseiva, P., & Taroni, F. (2014). Bayesian classification criterion for forensic multivariate data. *Forensic Science International, 244*, 295–301.

Bozza, S., Scherz, V., Greub, G., Falquet, L., & Taroni, F. (2022). A probabilistic approach to evaluate salivary microbiome in forensic science when the defense says: 'It is my twin brother'. *Forensic Science International:Genetics, 57*, 102638. https://doi.org/10.1016/j.fsigen.2021.102638

Broséus, J., Anglada, F., & Esseiva, P. (2010). The differentiation of fibre- and drug type cannabis seedling by gas chromatography/mass spectrometry and chemometric tools. *Forensic Science International, 200*, 87–92.

Brozek-Mucha, Z., & Jankowicz, A. (2001). Evaluation of the possibility of differentiation between various types of ammunition by means of GSR examination with SEM-EDX method. *Forensic Science International, 123*, 39–47.

Buckleton, J. S., Bright, J., & Taylor, D. E. (2016). *Forensic DNA Evidence Interpretation* (2nd ed.). Boca Raton: CRC Press.

Bunch, S. (2000). Consecutive matching striations criteria: A general critique. *Journal of Forensic Sciences, 45*, 955–962.

Cardinetti, B., Ciampini, C., Abate, S., Marchetti, C., Ferrari, F., Di Tullio, D., D'Onofrio, C., Orlando, G., Gravina, L., Torresi, L., & Saporita, G. (2006). A proposal for statistical evaluation of the detection of gunshot residues on a suspect. *Scanning, 28*, 142–147.

Casella, G., & Berger, R. L. (2002). *Statistical Inference* (2nd ed.). Pacific Grove: Duxbury Press.

Champod, C., Evett, I., & Jackson, G. (2004). Establishing the most appropriate databases for addressing source level propositions. *Science & Justice, 44*, 153–164.

Champod, C., Lennard, C., Margot, P., & Stoilovic, M. (2016). *Fingerprints and other ridge skin impressions* (2nd ed.). Boca Raton: CRC Press.

Chib, S. (1995). Marginal likelihood from the Gibbs output. *Journal of the American Statistical Association, 90*, 1313–1321.

Chib, S., & Jeliazkov, S. (2001). Marginal likelihood from the Metropolis-Hastings output. *Journal of the American Statistical Association, 96*(453), 270–281.

Cole, S. A. (2014). Forensic science and miscarriages of justice. In G. Bruinsma & D. Weisburd (eds.), *Encyclopedia of Criminology and Criminal Justice* (pp. 1763–1773). New York: Springer.

Cook, R., Evett, I. W., Jackson, G., Jones, P. J., & Lambert, J. A. (1998). A hierarchy of propositions: Deciding which level to address in casework. *Science & Justice, 38*, 231–239.

Cornfield, J. (1967). Bayes theorem. *Review of the International Statistical Institute, 35*, 34–49.

Cowell, R. G., Dawid, A. P., Lauritzen, S. L., & Spiegelhalter, D. J. (1999). *Probabilistic networks and expert systems*. New York: Springer.

D'Agostini, G. (2004). *Bayesian Reasoning in Data Analysis*. Singapore: World Scientific Publishing Co.

Davis, L., Saunders, C., Hepler, A., & Buscaglia, J. (2012). Using subsampling to estimate the strength of handwriting evidence via score-based likelihood ratios. *Forensic Science International, 216*, 146–157.

Dawid, A. P. (2017). Forensic likelihood ratio: Statistical problems and pitfalls. *Science & Justice, 57*, 73–75.

de Finetti, B. (1989). Probabilism. *Erkenntnis, 31*, 169–223.

de Finetti, B. (1993a). On the subjective meaning of probability (Paper originally published in the 'Fundamenta mathematicae', 17, 1931, pp. 298–329). In P. Monari & D. Cocchi (eds.), *Probabilità e induzione* (pp. 291–321). Bologna: CLUEB.

de Finetti, B. (1993b). Recent suggestions for the reconciliation of theories of probability (Paper originally published in the "Proceedings of the Second Berkely Symposium on Mathematical Statistics and Probability", held from July 31 to August 12, 1950, University of California Press, 1951, pp. 217–225). In P. Monari & D. Cocchi (eds.), *Probabilità e induzione* (pp. 375–387). Bologna: CLUEB.

de Finetti, B. (2017). *Theory of probability - A critical introductory treatment*. Chichester: Wiley.

Drygajlo, A., Jessen, M., Gfroerer, S., Wagner, I., Vermeulen, J., & Niemi, T. (2015). *Methodological guidelines for best practice in forensic semiautomatic and automatic speaker recognition (including guidance on the conduct of proficiency testing and collaborative exercises)*. http://enfsi.eu/wp-content/uploads/2016/09/guidelines_fasr_and_fsasr_0.pdf

Edwards, W. (1988). Insensitivity, commitment, belief and other Bayesian virtues, or, who put the snake in the warlord's bed? In P. Tillers & E. D. Green (eds.), *Probability and Inference in the Law of Evidence, The Uses and Limits of Bayesianism (Boston Studies in the Philosophy of Science)* (pp. 271–276). Dordrecht: Springer.

Evett, I. W. (1987). Bayesian inference and forensic science: problems and perspectives. *The Statistician, 36*, 99–105.

Evett, I. W. (1990). The theory of interpreting scientific transfer evidence. In *Forensic science progress* (Vol. 4, pp. 141–179). Berlin: Springer.

Evett, I. W. (1996). Expert evidence and forensic misconceptions of the nature of exact science. *Science & Justice, 36*, 118–122.

Evett, I. W., Jackson, G., Lambert, J. A., & McCrossan, S. (2000). The impact of the principles of evidence interpretation and the structure and content of statements. *Science & Justice, 40*, 233–239.

Franco-Pedroso, J., Ramos, D., & Gonzalez-Rodriguez, J. (2016). Gaussian mixture-models of between-source variation for likelihood ratio computation from multivariate data. *PLOS One, 11*, e0149958.

Friel, N., & Pettitt, A. N. (2008). Marginal likelihood estimation via power posteriors. *Journal of the Royal Statistical Society, Series B, 70*, 589–607.

Gaborini, L. (2019). *R package bayessource*. https://doi.org/10.5281/zenodo.3570578

Gaborini, L. (2021). *Bayesian Models in Questioned Handwriting and Signatures*. Ph.D. thesis, École des Sciences Criminelles, Université de Lausanne.

Gamerman, D., & Lopes, H. F. (2006). *Markov chain Monte Carlo: Stochastic simulation for Bayesian inference*. London: Chapman & Hall.

Garthwaite, P. H., & Al-Awadhi, S. A. (2001). Non-conjugate prior distribution assessment for multivariate normal sampling. *Journal of the Royal Statistical Society B, 63*, 95–110.

Garthwaite, P. H., Kadane, J. B., & O'Hagan, A. (2005). Statistical methods for eliciting probability distributions. *Journal of the American Statistical Association, 470*, 680–700.

Gelman, A., Carlin, J. B., Stern, H. S., Dunson, D., Vehtari, A., & Rubin, D. B. (2014). *Bayesian data analysis* (3rd ed.). Boca Raton: CRC Press.

Genz, A., Bretz, F., Miwa, T., Mi, X., Leisch, F., Scheipl, F., Bornkamp, B., Maechler, M., & Hothorn, T. (2020). *mvtnorm*. https://cran.r-project.org/web/packages/mvtnorm/mvtnorm.pdf

Geweke, J. (1989). Bayesian inference in econometric models using Monte Carlo integration. *Econometrica, 57*, 1317–1339.

Gill, P., Hicks, T., Butler, J. M., Connolly, E., Gusmão, L., Kokshoorn, B., Morling, N., van Oorschot, R., Parson, W., Prinz, M., Schneider, P. M., Sijen, T., & Taylor, D. (2018). DNA Commission of the International Society for Forensic Genetics: Assessing the value of forensic biological evidence – Guidelines highlighting the importance of propositions. Part I: Evaluation of of DNA profiling comparisons given (sub-) source propositions. *Forensic Science International: Genetics, 36*, 189–202.

Goldmann, T., Taroni, F., & Margot, P. (2004). Analysis of dyes in illicit pills (amphetamine and derivates). *Journal of Forensic Sciences, 49*, 716–722.

Good, I. J. (1950). *Probability and the weighting of the evidence*. London: Charles Griffin.

Good, I. J. (1958). Significance tests in parallel and in series. *Journal of the American Statistical Association, 53*, 799–813.

Good, I. J. (1988). The interface between statistics and philosophy of science. *Statistical Science, 4*, 386–397.

Hall, B., Hall, M., Statisticat, L., Brown, E., Hermanson, R., Charpentier, E., Heck, D., Laurent, S., Gronau, S. L., & Singmann, H. (2020). *Package 'LaplacesDemon'*. https://cran.r-project.org/web/packages/LaplacesDemon/LaplacesDemon.pdf

Han, C., & Carlin, B. (2001). Markov chain monte carlo methods for computing Bayes Factors: A comparative review. *Journal of the American Statistical Association, 96*, 1122–1132.

Hepler, A., Saunders, C., Davis, L., & Buscaglia, J. (2012). Score-based likelihood ratios for handwriting evidence. *Forensic Science International, 219*, 129–140.

Hopwood, A., Puch-Solis, R., Tucker, V., Curran, J., Skerrett, J., & Tully, G. (2012). Consideration of the probative value of single donor 15-plex STR profiles in UK courts. *Science & Justice, 52*, 185–190.

Howson, C., & Urbach, P. (1996). *Scientific reasoning: The Bayesian approach* (2nd edn.) Chicago: Open Court Publishing Company.

Jackson, G. (2000). The scientist and the scales of justice. *Science & Justice, 40*, 81–85.

Jackson, G., Jones, S., Booth, G., Champod, C., & Evett, I. W. (2006). The nature of forensic science opinion - A possible framework to guide thinking and practice in investigations and in court proceedings. *Science & Justice, 46*, 33–44.

Jacquet, M., & Champod, C. (2020). Automated face recognition in forensic science: Review and perspectives. *Forensic Science International, 307*, 110124.

Jeffrey, R. C. (1975). Probability and falsification: critique of the Popper program. *Synthese, 30*, 95–117.

Jeffreys, H. (1961). *Theory of probability* (3rd ed.). Oxford: Clarendon Press, Oxford.

Kamath, S., Charles Stephen, J. K., Suresh, S., Barai, B. K., Sahoo, A., Radhika Reddy, K., & Bhattacharya, K. R. (2008). Basmati rice: Its characteristics and identification. *Journal of the Science of Food and Agricolture, 88*, 1821–1831.

Kass, R. E. (1993). Bayes Factors in practice. *The Statistician, 42*, 551–560.

Kass, R. E., & Raftery, A. E. (1995). Bayes factors. *Journal of the American Statistical Association, 90*, 773–795.

Kaye, D. H. (2009). Trawling, DNA databases for partial matches: What is the FBI afraid of? *Cornell Journal of Law and Public Policy, 9*, 145–171.

Kruschke, J. K. (2015). *Doing Bayesian data analysis* (2nd ed.). London: Academic Press.

Lavine, M., & Schervish, M. J. (1999). Bayes factors: What they are and what they are not. *The American Statistician, 53*, 119–122.

Lee, P. M. (2012). *Bayesian statistics* (4th ed.). Chichester: Wiley.

Leegwater, A. J., Meuwly, D., Sjerps, M., Vergeer, P., & Alberink, I. (2017). Performance study of score-based likelihood ratio system for forensic fingermark comparison. *Journal of Forensic Sciences, 62*, 626–640.

Linden, J., Marquis, R., Bozza, S., & Taroni, F. (2018). Dynamic signatures: A review of dynamic feature variation and forensic methodology. *Forensic Science International, 291*, 216–229.

Linden, J., Taroni, F., Marquis, R., & Bozza, S. (2021). Bayesian multivariate models for case assessment in dynamic signature cases. *Forensic Science International, 318*, 110611.

Lindley, D. (2014). *Understanding uncertainty* (revised edition). Hoboken: Wiley.

Lindley, D. V. (1977). A problem in forensic science. *Biometrika, 64*, 207–213.

Lindley, D. V. (1985). *Making decisions* (2nd ed.). Chichester: Wiley.

Lindley, D. V. (2000). The philosophy of statistics. *The Statistician, 49*, 293–337.

Liu, C. C., & Aitkin, M. (2008). Bayes factors: Prior sensitivity and model generalizability. *Journal of Mathematical Psicology, 52*, 362–375.

Marin, J., & Robert, C. (2014). *Bayesian essentials with R* (2nd ed.). New York: Springer.

Marquis, R., Schmittbuhl, M., Mazzella, W., & Taroni, F. (2005). Quantification of the shape of handwritten characters loops. *Forensic Science International, 164*, 211–220.

Marquis, R., Taroni, F., Bozza, S., & Schmittbuhl, M. (2006). Quantitative characterization of morphological polymorphism of handwritten character loops. *Forensic Science International, 164*, 211–220.

Marquis, R., Biedermann, A., Cadola, L., Champod, C., Gueissaz, L., Massonnet, G., Mazzella, W., Taroni, F., & Hicks, T. (2016). Discussion on how to implement a verbal scale in a forensic laboratory: Benefits, pitfalls and suggestions to avoid misunderstanding. *Science & Justice, 56*, 364–370.

Martin, A. D., Quinn, K. M., Park, J. H., Vieilledent, G., Malecki, M., Blackwell, M., Poole, K., Reed, C., Goodrich, B., Ihaka, R., The R Development Core Team, The R Foundation, L'Ecuyer, P., Matsumoto, M., & Nishimura, T. (2021). *Package 'MCMCpack'*. https://cran.r-project.org/web/packages/MCMCpackr/MCMCpack.pdf

Meuwly, D. (2001). *Reconnaissance de locuteurs en sciences forensiques : l'apport d'une approche automatique.* Ph.D. thesis, Institut de Police Scientifique et de Criminologie, Université de Lausanne.

Meuwly, D., Ramos, D., & Haraksim, R. (2017). A guideline for the validation of likelihood ratio methods used for forensic evidence evaluation. *Forensic Science International, 276*, 142–153.

Morrison, G. S. (2016). Special issue on measuring and reporting the precision of forensic likelihood ratios: introduction to the debate. *Science & Justice, 56*, 371–373.

Morrison, G. S., Enzinger, E., Hughes, V., Jessen, M., Meuwly, D., Neumann, C., Planting, S., Thompson, W. C., van der Vloed, D., Ypma, R. J. F., Zhang, C., Anonymous, A., &

Anonymous, B. (2021). Consensus on validation of forensic voice comparison. *Science & Justice, 61*, 299–309.

Neal, R. (1996). *Bayesian learning for neural networks*. New York: Springer.

Neumann, C. (2020). Defence against the modern arts: the curse of statistics: Part I–FRStat. *Law, Probability and Risk, 19*, 1–20.

Neumann, C., & Ausdemore, M. (2020). Defence against the modern arts: The curse of statistics: Part II: 'score-based likelihood ratios'. *Law, Probability and Risk, 19*, 21–42.

Nic Daéid, N., Biedermann, A., Champod, C., Hutton, J., Jackson, G., Kitchin, D., Neocleous, T., Spiegelhalter, D., Willis, S., & Wilson, A. (2020). The use of statistics in legal proceedings: a primer for courts. Edinburgh: The Royal Society and The Royal Society of Edinburgh.

Nordgaard, A., Ansell, R., Drotz, W., & Jaeger, L. (2012). Scale of conclusions for the value of evidence. *Law, Probability and Risk, 11*, 1–24.

Oakley, J. (2008). *Package 'SHELF'*. https://cran.r-project.org/web/packages/SHELF/SHELF.pdf

O'Hagan, A., & Kendall, M. (1994). *Kendall's advanced theory of statistics: Bayesian inference. Volume 2B*. Number v. 2, pt. 2 in Kendall, Maurice George. Kendall's Advanced Theory of Statistics. Edward Arnold.

O'Hagan, A., Buck, C. E., Daneshkhah, A., Eiser, J. R., Garthwaite, P. H., Jenkinson, D. J., Oakley, J. E., & Rakow, T. (2006). *Uncertain judgements: Eliciting experts' probabilities*. Hoboken: Wiley.

Ommen, D., Saunders, P., & Neumann, C. (2017). The characterization of Monte Carlo errors for the quantification of the value of forensic evidence. *Journal of Statistical Computation and Simulation, 87*, 1608–1643.

Ommen, D. M., Saunders, C. P., & Neumann, C. (2016). An argument against presenting interval quantifications as a surrogate for the value of evidence. *Science & Justice, 56*, 383–387.

PCAST (2016). *Forensic science in criminal courts: Ensuring scientific validity of feature-comparison methods*. Executive Office of the President President's Council of Advisors on Science and Technology (PCAST), Washington, DC.

Press, S. J. (2003). *Subjective and objective Bayesian stastistics*. Hoboken: Wiley.

Press, S. J. (2005). *Applied multivariate analysis. Using Bayesian and frequentist methods of inference* (2nd ed.). New York: Dover Publications, Inc.

Ramos, D., & Gonzalez-Rodriguez, J. (2013). Reliable support: Measuring calibration of likelihood ratios. *Forensic Science International, 230*, 156–169.

Ramos, D., Meuwly, D., Haraksim, R., & Berger, C. E. H. (2021). Validation of forensic automatic likelihood ratio methods. In D. Banks, K. Kafadar, D. Kaye, & M. Tackett (eds.), *Handbook of forensic statistics* (pp. 143–163). Boca Raton: CRC Press.

Robert, C. P. (2001). *The Bayesian choice* (2nd ed.). New York: Springer.

Robert, C. P., & Casella, G. (2010). *Introducing Monte Carlo methods with R*. New York: Springer.

Robertson, B., & Vignaux, G. A. (1993). Probability - The logic of the law. *Oxford Journal of Legal Studies, 13*, 457–478.

Robertson, B., & Vignaux, G. A. (1995). *Interpreting evidence. Evaluating forensic science in the courtroom*. Chichester: Wiley.

Robertson, B., Vignaux, G. A., & Berger, C. E. H. (2016). *Interpreting evidence. Evaluating forensic science in the courtroom* (2nd ed.). Chichester: Wiley.

Scherz, V. (2021). *Microbiota Profiling: Forensic Application*. Ph.D. thesis, Univertsity of Lausanne, Institute of Microbiology.

Scherz, V., Bertelli, C., Bozza, S., Aeby, S., Opota, O., Falquet, L., Taroni, F., & Greub, G. (2021). *When the defense says: "No, it is my twin brother": A salivary microbiota-based identification of monozygotic twins*. Technical report, University of Lausanne, Institute of Microbiology.

Scott, D. W. (1992). *Multivariate density estimation*. New York: Wiley.

Silverman, B. W. (1986). *Density estimation*. London: Chapman & Hall.

Sinharay, S., & Stern, H. (2002). On the sensitivity of Bayes Factors to the prior distributions. *The American Statistician, 56*, 196–201.

Sprenger, J. (2016). Bayesianism vs. frequentism in statistical inference. In A. Hàjek & C. Hitchcock (eds.), *The Oxford handbook of probability and philosophy* (pp. 382–405). Oxford: Oxford University Press.

Stan Development Team (2021). *Stan modeling language users guide and reference manual.* http://mc-stan.org

Tanner, M. A. (1996). *Tools for statistical inference: methods for the exploration of posterior distributions and likelihood functions* (3rd ed.). New York: Springer.

Taroni, F., Bozza, S., Biedermann, A., Garbolino, G., & Aitken, C. G. G. (2010). *Data analysis in forensic science: A Bayesian decision perspective.* Chichester: Wiley.

Taroni, F., Bozza, S., Biedermann, A., & Aitken, C. G. G. (2016). Dismissal of the illusion of uncertainty in the assessment of a likelihood ratio. *Law, Probability & Risk, 15,* 1–16.

Taroni, F., Garbolino, P., Biedermann, A., Aitken, C. G. G., & Bozza, S. (2018). Reconciliation of subjective probabilities and frequencies in forensic science. *Law, Probability & Risk, 17,* 243–262.

Taroni, F., Garbolino, P., & Bozza, S. (2020). Coherently updating degrees of belief: Radical Probabilism, the generalization of Bayes' Theorem and its consequences on evidence evaluation. *Law, Probability & Risk, 19,* 293–316.

Taroni, F., Garbolino, P., Bozza, S., & Aitken, C. (2021a) The Bayes' factor: the coherent measure for hypothesis confirmation. *Law, Probability & Risk, 20,* 15–36.

Taroni, F., Bozza, S., & Biedermann, A. (2021b). Decision theory. In D. Banks, K. Kafadar, D. Kaye, & M. Tackett (eds.), *Handbook of forensic statistics* (pp. 103–130). Boca Raton: CRC Press.

Tully, G. (2021). *Forensic Science Regulator Codes of Practice and Conduct,* Development of Evaluative Opinions FSR-C-118 Issue 1. London.

van Leeuwen, D. A., & Brümmer, N. (2013). The distribution of calibrated likelihood ratios in speaker recognition. In *Interspeech 2013, 14th Annual Conference of the International Speech Communication Association,* Lyon (pp. 1619–1623).

Verzani, J. (2014). *Using R for introductory statistics* (2nd ed.). Boca Raton: CRC Press.

Willis, S., McKenna, L., McDermott, S., O'Donell, G., Barrett, A., Rasmusson, B., Nordgaard, A., Berger, C., Sjerps, M., Lucena-Molina, J., Zadora, G., Aitken, C., Lovelock, T., Lunt, L., Champod, C., Biedermann, A., Hicks, T., & Taroni, F. (2015). *ENFSI guideline for evaluative reporting in forensic science, Strengthening the evaluation of forensic results across Europe (STEOFRAE).* Dublin. https://enfsi.eu/wp-content/uploads/2016/09/m1_guideline.pdf

Wilson, A., Aitken, C., Sleeman, R., & Carter, J. (2014). The evaluation of evidence relating to traces of cocaine on banknotes. *Forensic Science International, 236,* 67–76.

Wolodzko, T. (2020). *Package 'extraDistr'.* https://cran.r-project.org/web/packages/extraDistr/extraDistr.pdf

Zadora, G., Martyna, A., Ramos-Castro, D., & Aitken, C. G. G. (2014). *Statistical analysis in forensic science. Evidential value of multivariate physicochemical data.* Chichester: Wiley. Supplementary material (R codes and data) available at www.wiley.com/go/physicochemical

Zapata-Vazquez, R., O'Hagan, A., & Bastos, L. (2014). Eliciting expert judgements about a set of proportions. *Journal of Applied Statistics, 41,* 1919–1933.

Index

Printed in the United States
by Baker & Taylor Publisher Services